"This is a book that speaks to a great nee~~~ ~~ ~~ ~~~~~, ~~~ ~~~~ ~~~ depend on illusion. It takes very seriously both science and exegesis. It has the potential to mobilize authentic Christian hope in fearful and perplexing times."

Richard Bauckham, FBA, FRSE, professor emeritus, University of St Andrews

"*Let Creation Rejoice* is an extremely valuable addition to the literature and theology of creation care. Where evangelical Christianity has tended to see in eschatology an excuse for avoiding environmental questions altogether, White and Moo take the opposite approach. It is in the hope that is in Jesus' promised return to restore all things that they see our biggest reason for acting now. Yes, we live in an age of potential despair—but what hope there is in these pages. The honest, careful and encouraging exegesis here should bless all who read it."

Rev. Ed Brown, Care of Creation Inc.

"*Let Creation Rejoice* is a marvelously lucid study of Scripture and current ecological concerns. Those skeptical about the magnitude of the threats facing the earth will be challenged by the sober, balanced account of climate change and related issues; those skeptical of Christianity's relevance to such concerns will equally be challenged by the elegant exposition of the Bible's affirmation of the created order. I cannot imagine a more clearheaded and timely treatment of the topic."

Sean McDonough, professor of New Testament, Gordon-Conwell Theological Seminary

"If your mailboxes, physical and digital, are like mine, they are filled with apocalyptic rantings. Religious groups point to the signs of the end times. Environmentalist groups warn us we must act now, before it's too late. Here's a different message. While not every reader will agree with every perspective in this book, all of us can be provoked by its winsome engagement with tough issues from a biblical perspective. Whatever your position on matters such as climate change, this book will prompt you to think through how your biblical hope intersects with the problems of the day."

Russell D. Moore, president, Ethics & Religious Liberty Commission, Southern Baptist Convention

"The authors give a careful and comprehensive scientific analysis of contemporary threats to the environment, such as those posed by climate change. To this they add an extensive discussion of biblical material relating to human hope and responsibility. Many Christians will find this book a helpful approach to serious problems of our age."

Rev. Dr. John Polkinghorne, KBE, FRS

"This book shows why The Cape Town Commitment, from the Third Lausanne Congress, Cape Town 2010, was absolutely right to say that 'we cannot separate our relationship to Christ from how we act in relation to the earth. For to proclaim the gospel that says "Jesus is Lord" is to proclaim the gospel that includes the earth, since Christ's Lordship is over all creation. Creation care is thus a gospel issue within the Lordship of Christ.' To be able to make such an affirmation requires that we have a fully biblical vision of what the gospel wholly and actually is, and this book richly provides such a vision with extensive biblical support. It takes us through the whole Bible story, from creation to new creation, and shows how our attitudes and actions in relation to the earth must be shaped by what God has done and will do for all creation through the Lord Jesus Christ. The book presents and explains the sober facts of our present ecological crisis. But, as the title makes clear, it does so not to induce despair, but rather to show how radical and transformative is the hope for all creation that is truly part of the 'good news' of the biblical gospel."
Christopher J. H. Wright, Langham Partnership; author of *The Mission of God* and *The Mission of God's People*

"This powerful book is full of godly wisdom. Jonathan Moo and Robert White have given us an authoritative account of the current scientific data together with an inspirational review of a wide range of profound biblical resources. It is truly important reading for anyone seeking for hope in our troubled environmental times."
Peter Harris, A Rocha

"What a wonderful book. There isn't anything quite like it. Biblical scholar Moo and natural scientist White join hands to pen this compelling book on hope—authentic biblical hope—in a time of increasing despair. This book is both an honest and lucid presentation of the ecological challenges before us and an insightful and articulate discussion of important biblical texts such as Romans 8, 2 Peter 3 and Luke 12. I found myself again and again saying, 'That is exactly right' and 'Well said.' Perhaps the highest compliment I can give is that I intend to use this book in teaching my college and seminary classes. Take up and read."
Steven Bouma-Prediger, Hope College, author of *For the Beauty of the Earth*

LET CREATION REJOICE

BIBLICAL HOPE AND
ECOLOGICAL CRISIS

◆

JONATHAN A. MOO
and ROBERT S. WHITE

An imprint of InterVarsity Press
Downers Grove, Illinois

InterVarsity Press
P.O. Box 1400, Downers Grove, IL 60515-1426
Internet: www.ivpress.com
Email: email@ivpress.com

©2014 by Jonathan A. Moo and Robert S. White

Published in the United States of America by InterVarsity Press, Downers Grove, Illinois, with permission from Inter-Varsity Press, Leicester, England.

All rights reserved. No part of this book may be reproduced in any form without written permission from InterVarsity Press.

InterVarsity Press® is the book-publishing division of InterVarsity Christian Fellowship/USA®, a movement of students and faculty active on campus at hundreds of universities, colleges and schools of nursing in the United States of America, and a member movement of the International Fellowship of Evangelical Students. For information about local and regional activities, write Public Relations Dept., InterVarsity Christian Fellowship/USA, 6400 Schroeder Rd., P.O. Box 7895, Madison, WI 53707-7895, or visit the IVCF website at www.intervarsity.org.

All Scripture quotations, unless otherwise indicated, are taken from THE HOLY BIBLE, NEW INTERNATIONAL VERSION®, NIV® Copyright © 1973, 1978, 1984, 2011 by Biblica, Inc.™ Used by permission. All rights reserved worldwide.

Cover design: David Fassett
Interior design: Beth Hagenberg
Images: paper background: © tomograf/iStockphoto
  tree illustration: © John_ Woodcock/iStockphoto
  birds: © dbencek/iStockphoto

ISBN 978-0-8308-4052-6 (print)
ISBN 978-0-8308-9635-6 (digital)

Printed in the United States of America ∞

 InterVarsity Press is committed to protecting the environment and to the responsible use of natural resources. As a member of Green Press Initiative we use recycled paper whenever possible. To learn more about the Green Press Initiative, visit www.greenpressinitiative.org.

Library of Congress Cataloging-in-Publication Data
A catalog record for this book is available from the Library of Congress.

| P | 23 | 22 | 21 | 20 | 19 | 18 | 17 | 16 | 15 | 14 | 13 | 12 | 11 | 10 | 9 | 8 | 7 | 6 | 5 | 4 | 3 | 2 | 1 |
| Y | 33 | 32 | 31 | 30 | 29 | 28 | 27 | 26 | 25 | 24 | 23 | 22 | 21 | 20 | 19 | 18 | 17 | 16 | 15 | 14 |

Contents

Preface. 7

1. Apocalypse Now? Living in the Last Days 11

2. Life on Earth Today. 21

3. Global Climate Change 54

4. Why Hope? The Gospel and the Future 80

5. Bringing New Testament Hope Down to Earth 97

6. Cosmic Catastrophe?. 115

7. Jesus, a Thief in the Night and the Kingdom of God. 131

8. Revelation and the Renewal of All Things. 146

9. Finding Joy in an Active and Living Hope 162

Afterword: *Practical Resources* 171

Notes. 175

Scripture Index 185

Preface

◆

As we compose this preface, we are nearing the close of another year, and in its mix of joys and sorrows, in its hopes and disappointments, it has been a year much like any other—for us and for all of life on earth. Yet in writing this book we have been forced to confront the reality that there are also some things about this year, and indeed about every year now, that are unprecedented in the long history of human habitation of this marvelous planet. To name but one, Typhoon Haiyan hit the Philippines with record high wind speeds in excess of any previously recorded on land. Despite advance warning, it killed thousands and destroyed or damaged the homes of more than 16 million people. It is the worst natural disaster ever to affect the Philippines.

Droughts, heat waves, water shortages, forest fires, hurricanes, floods, the destruction of natural habitats, rapid decline in the abundance and diversity of plant and animal life—such things, we know, are not unique to our time. Many times in the history of our planet the intensity and severity of such events have in fact exceeded anything we experience today. A volcano erupts and wreaks destruction across an area the size of the United States, temporarily changing the weather across the entire globe; an asteroid hits the planet and causes the extinction of untold numbers of species; the earth wobbles in its orbit and the climate cools, sending miles-thick ice scouring the earth across continent-size land masses; thousands of years later, the climate warms, unleashing floods of unimaginable size that destroy all in their path and shape the landscape for eons to come. Such massive events and large-scale processes humble us; they reveal to us the forces that have shaped our wondrous, fruitful planet; and they caution us against extravagant claims about the uniqueness of our own age.

Yet what is inescapably different about today is that never in the history of human life have so many people been so threatened by the changes our planet is undergoing; never have some of the planetary changes we are witnessing occurred so quickly, with so little time for adaptation; and never before has one species (us) been identified as the primary cause of such rapid, large-scale changes. It is this recognition of our vulnerability and our culpability, along with the fear that things are on the verge of getting much, much worse and there is little we can do about it, that lies behind the despair so prevalent in this age. We increasingly observe the temptation to such despair among scientists, environmentalists, those who work for development and aid agencies, and even portions of the general public. In our own work and ministry—and, indeed, as we did research for the science portions of this book—we have occasionally wrestled with such despair ourselves.

This book, though, is about hope—and about the only and ultimate source of our hope. We are writing this preface not only toward the end of another year in the secular calendar, but during the Christian season of Advent. It is a time of darkness and of waiting for the light. We yearn for the coming of God in Christ, for the coming of our Creator, Savior and Lord, and for the light and life of the resurrection and new creation. At the end of this season we celebrate the answer to this longing: we celebrate the coming of God himself as the baby Jesus on the first Christmas Day. We witness in the birth of Jesus—and in his life, his suffering and his death—that God is truly with us in our weakness, in our suffering and in our sorrow; and we find in Jesus' resurrection the sure and certain hope of his coming again, of the resurrection and of the renewal of all things. In this hope, and through the life-giving power of the Holy Spirit, we are enabled to face the messed-up, broken world in which we live. We are challenged to go on faithfully living out and proclaiming the gospel—the good news about Jesus—even in an age of despair. The scope of this good news, of what God has done and will do in Christ, exceeds the capacity of our imaginations. Yet we have become convinced that, however limited our ability to envision our future and the future of life on earth, our age needs more than ever for the church to regain its confidence not only in the personal saving grace of God in Christ, but also in the renewal of the whole of the created order that the Bible tells us has been made possible through Christ's incarnation, death and resurrection.

As we say in the first chapter, it is our desire that readers come away from this book with a renewed appreciation of the wonderful world that God has created, as well as a firm understanding of its present condition and the potential that we have to affect it. But most of all we aim to encourage a profound trust in the Creator and Redeemer God whose faithfulness is the only and ultimate ground of our hope.

In our work on this book we have been greatly helped by the support we have received from the Faraday Institute for Science and Religion based in St Edmund's College, Cambridge, and from the Kirby Laing Institute for Christian Ethics at Tyndale House, also in Cambridge. We are grateful to InterVarsity Press and especially to Andy Le Peau, as well as Philip Duce of IVP-UK, for their close attention to our manuscript and insightful comments. Jonathan is grateful for the support of Whitworth University's Faculty Faith and Scholarship Program, for the many students, lecture audiences and church groups who have interacted with some of the material presented here and for the unstinting support of his colleagues in the Whitworth University theology department. He would like especially to thank Jim Edwards, Adam Neder and Jerry Sittser for reading portions of this book and providing invaluable advice. Jonathan also thanks his parents, Jenny and Doug Moo, for their ongoing support and for reading and commenting on portions of the book. We are especially grateful to Robert Nowak, who provided critical, detailed comments on several chapters that helped make the book better than it would have been otherwise. Sam Berry also read several chapters and provided us with some keen advice at a critical point. Roger Abbot, Evan Durrant and Stacey Moo each read the entire manuscript at a late stage, and we are very thankful for their insightful suggestions and comments. Near the end of this project, Julian

Hardyman (senior pastor of Eden Baptist Church, Cambridge) provided helpful suggestions and advice. And at the very beginning of this project, Mark Ashton (then vicar of the Round Church at St Andrew the Great, Cambridge) challenged us to always keep our focus on the gospel, a challenge that we have both taken to heart. We only wish he were still alive to tell us whether he would count our efforts here a success.

Finally, we are thankful above all to our wives, Stacey Moo and Helen White, for their partnership, love and support.

Jonathan Moo and Robert White
Advent 2013

1

Apocalypse Now?
Living in the Last Days

*It's the end of the world as we know it
(and I feel fine).*

R.E.M.

◆

CELEBRATING THE END

North Idaho is a thinly settled region of forests, lakes and mountains narrowly separating the states of Washington and Montana in the northwestern United States. Known mostly for logging, mining and outdoor recreation, it is a beautiful place and I (Jonathan) am grateful to live not far away in Spokane, Washington. But the wilds of North Idaho have an unfortunate reputation for attracting radical sects and fringe groups of all sorts, ranging from religious cults to militias to white supremacists. Most infamously, it was the site of the 1992 Ruby Ridge siege, where the combination of an apocalyptically minded family fearful of government conspiracy and heavily armed federal agents led to a shootout and several deaths. The stereotype is sufficiently strong that when I came across a *New York Times Magazine* article provocatively titled "The End Is Near (Yay!)" and featuring a group based in the North Idaho town of Sandpoint, I assumed that I was about to read about yet another religious cult heralding the impending end of the world.[1]

As it turned out, however, the group highlighted in the article is not nearly so far from the mainstream as you might expect. It does indeed anticipate the end of civilization as we know it, but it is not a narrow religious sect—or even religious at all in the traditional sense—and its beliefs about the future mirror those of many in our society. The movement of which it is a part, the Transition Network, was started by a teacher named Rob Hopkins in Cornwall, England. It has followers in towns and cities all around the world, including places like Sandpoint, Idaho, but also Bologna, Italy; Los Angeles, California; São Paulo, Brazil; Houston, Texas; Tokyo, Japan; Portland, Maine; Sydney, Australia; and Cambridge, England.[2] The mix of science, economics, politics and social analysis that grabs the attention of its followers is regularly the subject of newspaper and magazine articles and features in popular books and films. But what is perhaps most worrying for those of us tempted to dismiss groups like the Transition Network is that many of today's leading scientists share their rather bleak assessment of our civilization's future. When no less a figure than Martin Rees (formerly president of the Royal Society, the United Kingdom's national academy of science) suggests that there is only a 50 percent chance that civilization as we know it will make it through the present century,[3] it is probably time that the rest of us sat up and took notice.

This book was written to help readers understand why scientists are making such apparently radical claims. More importantly, it was written to help readers consider how Christians might respond if we take seriously what the Bible has to say about the gospel and the future of life on earth. In a world increasingly tempted to despair in the face of dysfunctional politics and economic and ecological crises, we need now more than ever to be able to articulate clearly the hope we have in Christ—and to embody that hope in our lives and actions.

Compared to some other recent movements, the Transition Network actually adopts a relatively optimistic stance toward the future: "engaged optimism" is how its founder, Hopkins, describes it. Its followers embrace the changes that are thought necessary to "transition" from our present unsustainable way of life to a more sustainable one as they seek to mitigate the effects of climate change and to strengthen the resilience of local communities. There are many other thinkers and scientists, however, who are much more pessimistic about the future, and their numbers are growing.

Paul Kingsnorth, leader of a UK-based group calling itself Dark Mountain, claims that there is no longer any hope that we can prevent catastrophic climate change and environmental collapse. He urges us essentially to give up on the present and to focus on getting ready for the sort of world that might emerge after the environmental apocalypse he expects in the relatively near future. Dark Mountain garnered publicity shortly after its founding in 2009 when the *Guardian* newspaper published a debate between Kingsnorth and the well-known political and environmental journalist George Monbiot.[4] Monbiot argued strongly against what he saw as Kingsnorth's defeatism, observing that movements like Dark Mountain serve only to allow those who exploit the earth for their own ends to go on doing so and thus to hasten the realization of their own apocalyptic predictions (predictions that, Monbiot pointed out, would see billions of people condemned to a ruinous future). But what was most striking about the debate between Monbiot and Kingsnorth was the extent to which they basically agreed about how the future is likely to turn out. Although they call for very different responses to the threats facing life on earth—and Monbiot clings tenaciously to the hope that we might still avert catastrophe (the hope that also motivates Hopkins and his Transition Network)—even Monbiot concedes that, given the way things are going, the outlook for our future is dark and foreboding.

A Perfect Storm, and Other Forecasts

The notion that a so-called perfect storm of factors are coming together in a way that threatens the future of life on earth is no longer the unique preserve of bearded prophets, street preachers and religious fundamentalists. John Beddington, the United Kingdom's chief scientist, provoked debate a few years ago when he suggested that we might begin witnessing the catastrophic effects of just such a perfect storm as soon as 2030. The accusations of scaremongering that ensued suggest that 2030 was perhaps just a little too close for comfort even for a public that has grown used to terrifying predictions about events fifty or a hundred years down the road.[5]

The natural history presenter David Attenborough recently admitted that, in his view, "the world is in terrible trouble.... Am I optimistic about the future? No, not at all. But that's irrelevant. It's imperative that you do something, even if you don't think it's going to do any good."[6]

The surest sign of the popularity of apocalyptic rhetoric about the environment, however, is that even politicians occasionally get into the act. In the run-up to the Copenhagen Summit on climate change in late 2009, the UK's then prime minister, Gordon Brown, warned of the danger of impending "climate catastrophe."[7] And of course former US vice president Al Gore's controversial film *An Inconvenient Truth* frightened plenty of people with its computer-generated images of rising sea levels inundating New York City (despite the fact that few scientists expect anything like that to happen for centuries), and unfortunately also further politicized what was already a deeply polarizing issue.

The film industry unsurprisingly has taken advantage of the cinematic potential of our fears about the future. A few movies explicitly reflect contemporary concerns about the environment or climate change (*The Day After Tomorrow*, 2004; *The Age of Stupid*, 2009), whereas many more invent their own apocalyptic scenarios (*Children of Men*, 2006; *I Am Legend*, 2007; *The Book of Eli*, 2010; *Melancholia*, 2011), or leave undetermined the causes of civilization's end (*The Road*, 2009, based on Cormac McCarthy's haunting novel of the same name). A recent National Geographic reality television show in the United States, *Doomsday Preppers,* reveals just how serious are some people's fears by featuring a variety of Americans preparing for disaster and the breakdown of civilization that they expect in the not-so-distant future. (Their preparations generally seem to involve growing their own produce and storing food, water and lots of guns and ammunition.) Meanwhile, in the more rarefied realm of philosophy, the popular Slovenian philosopher and cultural critic Slavoj Žižek has written a book about our age appropriately titled *Living in the End Times.*[8]

Sociologists can no doubt come up with their own explanations for the popularity of this recent talk of impending apocalypse; such language has been around for millennia, even if the causes of the predicted catastrophes have varied.[9] What we are concerned with in this book, however, is first to assess the scientific data that provides the fuel upon which today's rhetoric burns. Do all of these dire predictions, as many pundits suggest, amount to nothing more than ideological scaremongering, perhaps hyped up for political or personal ends? Or are there good reasons for thinking that we may indeed be facing a crisis unprecedented in its scale and in the severity of its effects on life on earth?

We encourage readers to assess the evidence for themselves. We have tried to help by summarizing what seems to us to be the most relevant data in the following two chapters. As you will discover, our own assessment leads us to conclude that there is in fact plenty of cause for concern—and that is part of the reason why we have written this book. Climate change, we will suggest, is only the most publicized (and, admittedly, potentially the most far-reaching) threat that our planet faces in the coming decades. There is a wide range of much more obvious, interrelated and damaging effects that an ever-growing number of people consuming more and more are having on the planet on which we all depend.

To give you a quick idea of the sort of things that we have in mind, here is one way of understanding our situation as it has been summarized recently in *Nature*, one of the world's leading scientific journals:

> [A] group of leading academics argue that humanity must stay within defined boundaries for a range of essential Earth-system processes to avoid catastrophic environmental change. . . . They propose that for three of these—the nitrogen cycle, the rate of loss of species and anthropogenic climate change—the maximum acceptable limit has already been transgressed. In addition, they say that humanity is fast approaching the boundaries for freshwater use, for converting forests and other natural ecosystems to cropland and urban areas, and for acidification of the oceans. Crossing even one of these planetary boundaries would risk triggering abrupt or irreversible environmental changes that would be very damaging or even catastrophic for society.[10]

Consider what these scientists are claiming: crossing even one of these so-called planetary boundaries "would risk triggering abrupt or irreversible environmental changes that would be very damaging or even catastrophic for society"—and we have already crossed three of them and are rapidly approaching three more. The basis for such claims and their reliability are discussed in the following chapter, and there you will also have the chance to examine the hard data more closely for yourself. We urge readers to take some time to consider what many of the best biologists, earth scientists and climatologists have to report to us as they monitor and study our incredible planet. We recognize that many of us have grown cynical and weary in the face of the deluge of apocalyptic rhetoric with which politicians, the press and all manner of special interest groups inundate us. Nevertheless, it will not do for us simply to ignore what is going on in the world around us. Above all, as we will argue later in this book, Christians have a particular responsibility to seek to understand the world in which we live.

How to Respond?

If, as we ourselves have been forced to conclude, consideration of the scientific data does give cause for serious concern, that is not to say that the next step for us all is simply to jump on the bandwagon of secular environmentalists or "climate change Cassandras" (as some activists have recently been called). As Christians, our vision of the future (and hence our view of the present too) is necessarily going to be different from those who do not share our biblical hope. Our task in the second half of this book is to reflect on just what difference it makes to how we respond to the environmental challenges facing us if we take seriously the picture of the future that Scripture paints for us. What does the Bible say about the future of the earth, and what difference does that make to how we live now? What should be distinctive about how Christians engage with the sort of issues that are summarized in the next two chapters?

Most of us, Christian or not, have probably not entirely bought into the apocalyptic rhetoric that marks many of today's discussions of climate change and the environment. Many of us, however, might find ourselves vaguely identifying with one or more of the categories of response that we catalog below. Do you find yourself in any of these groups?

Ignorance-is-blissers: The issues concerning environmental degradation, resource depletion, biodiversity loss and climate change are complicated and heavily disputed; many of us therefore simply cannot be bothered to try to understand them, or we find it too difficult and time consuming to assess the evidence for ourselves. The issues can also seem irrelevant to our day-to-day lives, and so—we think—they are best left to others.

Seekers: This group have a sincere interest in the issues but simply do not know what to make of all the debates over climate change and the environment. They need to be convinced that concerns about such things are well founded and based on facts before they will consider taking any action. Many in the United States in particular are suspicious of the motivations behind the relevant scientific research, and the media has often fed such suspicions (partly because of the effective work of the next group) and left them uncertain of what to make of it all.

Deniers: It is possible to be convinced to some degree about the severity of a threat but not be prepared seriously to face the consequences or be willing to do anything about it—especially when we think it might involve some sort of personal sacrifice. Our denial of the reality or the significance of environmental challenges often stems simply from our lack of clear information or our lack of interest in spending time worrying about it. But in a few cases, as in the well-documented campaign by a handful of oil companies to discredit climate science (a campaign deliberately modeled on the tobacco industry's attempt to discredit medical findings about the effects of tobacco use),[11] denial stems from an unwillingness to contemplate changing the way we do things and a blatant desire to prevent any new information from coming to light that might get in the way of business as usual.

Problem solvers: This group represent quite a spectrum, from those who are convinced that there are quick, relatively painless technological fixes to most of the challenges that we face today, to those who think that radical economic and social restructuring is necessary if we are to prevent environmental catastrophe. Some in this group are progressive optimists who have faith in the ingenuity and potential of humankind; others might find themselves nearer despair about the future but nonetheless get on with trying to change things for the better.

Despairers: People who have despaired about the future of life on earth are often those who have spent years trying to change things but have seen little progress. Some are already suggesting that it is too late to prevent catastrophic climate change or catastrophic environmental collapse of one sort or another and that we would be better off at this point to focus our money and our efforts on adaptation.

Post-apocalypse hopers: The Dark Mountain group would fit in this category. For people in this group, collapse is inevitable, but they are intrigued by what might emerge afterward—perhaps a smaller human community living more sustainably on the earth or, in radical versions, an earth without human beings at all (the sort of future envisioned in Alan Weisman's creative thought experiment *The World Without Us*).[12] There are few people who are likely to find themselves in this category at the moment, but it may grow in popularity if things in coming years do indeed begin to look as bad as some predict. There is a rather obvious parallel between this secular hope and certain popular versions of Christian hope, even if the nature of the "apocalypse" and the ensuing future age envisaged are quite different.

As you might have guessed, we do not find any of these categories adequate for capturing what a distinctively Christian response should look like—although we do hope that a biblical perspective on the future will prompt many of us to enter into the difficult tasks taken on by those whom we have labeled "problem solvers." But if the Christian gospel fundamentally reorients us in our relationship with God and his world, then there ought to be something fundamentally different in our approach and in our attitude toward how we engage with our fellow human beings and with the rest of creation.

Looking Ahead

It might surprise some readers that we have decided to focus on what the Bible says about the future rather than on, for example, the doctrine of creation or the scope of Christian mission or the implications of loving God and neighbor. These indeed can all serve as compelling, biblically rooted starting points in a discussion of how Christians might respond to environmental challenges. But—as is amply attested by some of the responses to environmental issues that we have summarized above—our view of the future can and does have a profound effect on how we engage with the present.

This is no less true for Christians than it is for anyone else, and Christians have sometimes been accused precisely of letting our view of the future world affect our involvement (or lack of involvement) in the present world. The biblical view of creation is in any case directed toward the future promised us in Christ, and for Christians *creation* can hardly be considered apart from *new creation*. The scope and contours of biblical hope have important implications for how we express Christian love and charity. "The greatest of these is love," the apostle Paul reminds us (1 Cor 13:13); but of course he also tells us of the necessity of faith and hope for Christian living now.

In chapter 4 we discuss at greater length the importance of thinking carefully about what Scripture teaches about the future—and in particular about the future of the nonhuman creation. We also deal there with some criticisms of Christian eschatology (the doctrine of the "last things") that have been made by those who find it more of a hindrance than a help when addressing environmental issues. Before we get there, however, the next two chapters attempt to summarize the results of an up-to-date "health check" of our planet. What is the status of life on earth today according to the best insights that science can offer us? What is its future prognosis?

Chapter 2 treats a wide range of factors that are critical for the functioning of the earth's ecosystems and for human life and well-being, many of them seemingly mundane and straightforward (e.g., the availability of fresh water), some of them probably less obvious to most of us (e.g., the nitrogen cycle). Climate change is examined in chapter 3. A separate chapter is called for because of the complexity of the subject, its common misrepresentation in the popular press and its potential to have significant effects on many of the other systems and processes treated in chapter 2. Chapter 3 tackles the following questions: What is the evidence for human-influenced climate change? What is the likely trajectory of temperature changes in the future? How certain can we be about any of it, does it matter and is there anything that can be done to change the course we are on?

After our defense in chapter 4 of why we think it is important to ask what the Bible teaches about the future, we work carefully through a number of key biblical passages in chapters 5 through 8 in order to gain some insight into just what the Christian Scriptures have to tell us about God's plan for the earth and how this fits into the story of what he has accomplished for us in Christ. Chapter 9 then draws out some of the implications—if we are to take seriously what the Bible says about the future—for living on earth now, for living in a world that faces the sorts of challenges outlined in chapters 2 and 3.

It is our desire that readers come away from this book with a renewed appreciation of the wonderful world that God has created, as well as a firm understanding of its present condition and the potential that we have to affect it. But most of all we aim to encourage a profound trust in the Creator and Redeemer God whose faithfulness is the only and ultimate ground of our hope.

2

Life on Earth Today

*Even in a cosmic or a geological time-perspective,
there's something unique about our century:
for the first time in its history, our entire
planet's fate depends on human
actions and human choices.*

MARTIN REES,
PRESIDENT OF THE ROYAL SOCIETY

*For you make me glad by your deeds, LORD;
I sing for joy at what your hands have done.
How great are your works, LORD,
how profound your thoughts!*

PSALM 92:4-5

◆

THE PALE BLUE DOT

One of the most striking images of Earth was taken in 1968 from the lunar module of *Apollo 8* as it orbited the moon. The photo shows a pearly blue globe against a pitch-black sky as the earth rises above the barren surface of the moon in the foreground. It has become humankind's most reproduced image, ever. It epitomizes the beauty and the fragility of this tiny planet on which we all live, floating in the immensity of space.

The Old Testament psalmist, writing millennia before our time, has a surprisingly similar vision of the beauty and vastness of a universe within which humankind appears small and insignificant. As David sat caring for his sheep in the darkness of a Middle Eastern hillside long before the era of artificial light pollution, he looked out in the other direction, from earth to space. With a deep sense of awe and wonder David[1] wrote in Psalm 8 of his sovereign Lord, who had set his "glory in the heavens":

When I consider your heavens,
> the work of your fingers,
the moon and the stars,
> which you have set in place,
what is mankind that you are mindful of them,
> human beings that you care for them?

And yet, he went on to write:

You have made them a little lower than the angels
> and crowned them with glory and honor. (Ps 8:1, 3-5)

The astonishing truth we face as we consider the place of humankind on this planet is that, as the psalmist goes on to say, God has given us dominion over the works of his hands. Human beings, with all their frailty, have been given the responsibility by God to care for the amazingly fruitful world which he has created.

If we were not already aware of the lonely isolation of our home in the immensity of the cosmos, an image taken by the *Voyager* spacecraft in 1990 from the edge of the solar system 3.8 billion miles (6 billion kilometers) away reinforced it a million-fold. The earth is a tiny, pale blue dot, "a lonely speck in the great enveloping cosmic dark," as the writer Carl Sagan put it. Sagan went on to observe that "the Earth is the only world known so far to harbor life. There is nowhere else, at least in the near future, to which our species could migrate. Visit, yes. Settle, not yet. Like it or not, for the moment the Earth is where we make our stand."[2]

Beautiful yet fragile. All of human life and history—its triumphs and catastrophes, its joys and sorrows, its heroes and villains—has been played out in the tiny layer of the biosphere that envelopes the solid earth.

Seen from space, if we imagine the earth as the size of a billiard ball, we humans live entirely on a sheath around the surface that is not much more than the thickness of the trace of a fingerprint that would be left by picking the ball up.

We are all dependent on this thin carapace around this one planet for our well-being and indeed for our very lives. For now and for the foreseeable future we only have one planet on which to live. Yet, as we will see in this chapter, those of us who live in the world's high-income countries are consuming resources at a rate that would require far more than one planet Earth to sustain. In terms of the earth's biocapacity, those living in the United Kingdom are using nearly three times the ecological capacity per capita of the earth as a whole; in the United States the excess usage is even higher, more than four times the earth's capacity.[3] Some of the earth's resources we have in fact already used up or destroyed, so they will never again be available to those who follow us. We have, knowingly or not, annihilated thousands of living species and continue to do so at a shocking rate. We are polluting the land, sea and air, with little understanding of what might be the long-term consequences.

The changes taking place in the world around us are happening alarmingly fast from a biological or geological perspective. Yet by and large, most of us do not really notice them; they appear slow or imperceptible when viewed from the standpoint of our busy lives. There are often seemingly more pressing issues at hand, whether they are in the immediate orbit of our work, families and local communities or on the national or international political stage. It is hard to focus on changes to the environment that happen over decades or centuries when we each live only a few score years. Individual politicians usually exert power for a relatively short period and, in democracies, ruling parties are elected for only three or four years at a time. So it is especially difficult for politicians to focus on longer-term issues or to implement costly changes that will not bear fruit for many years or decades.

There are two main reasons, then, why it is profoundly difficult for us to take environmental change seriously. The first, as we have just observed, is that it often occurs relatively slowly in comparison to human time frames.

As marshes were drained, rivers dammed and diverted, and woods and hedges of our great-grandparents' time gradually cleared for agriculture and development, each successive generation started from a base where the marshes and forests were already smaller in extent; where the rivers had never run free and full of salmon; where species once locally common had always seemed rare or had been forgotten entirely. That was all they had known, so the further changes that occurred during their own lifetimes appeared less drastic. Maybe a few of those in the older generation grumbled that things were not what they had been in their youth, but then people always say that, don't they? And in the not-so-distant past, when life expectancies were on average considerably lower than today, there was even less time to notice the changes happening.

Jan Boersema has argued that even one of the icons of environmental catastrophe, the "collapse" of civilization on Easter Island, was probably not nearly so abrupt or catastrophic as many assume.[4] It is true that the island was gradually deforested, but the changes occurred slowly. The result was not a sudden collapse, but rather a slow decline in the wealth, variety of food and cultural richness of the society—a decrease in the quality of life. This created a gradual sociocultural transition, of which the inhabitants themselves may not even have been particularly aware while they were in the midst of it. Much the same might be said of the environmental changes happening today. In retrospect the major changes we are causing will no doubt be obvious (as with the deforestation of Easter Island), but they are far less obvious for many of us living through them. This phenomenon ought to warn us of the very real risk of sleepwalking into a situation where we have unknowingly degraded our natural environment to such an extent that its former richness—and the benefits it provides for human life and flourishing—is lost forever.

The second reason why we may find it difficult to recognize the magnitude of global environmental change is that most people, at least until the late twentieth century, lived out their lives in a very small geographical area. For example, the grandmother of Robert White's wife, who died at age ninety-three in 1989, never in her entire life travelled more than thirty miles from her home in the west of Britain. Her son has lived his whole life in only three houses, all within a mile of each other in the same town. For many in the world today, the reality is that they too will live out their entire lives in the same village or locality. So global changes, say in the melting of the Arctic ice cap, the rising of sea levels, the loss of forests or topsoil and the disappearance of species, are rarely likely to impinge greatly on their immediate awareness—until it is too late.

In the past few decades, however, technological advances have made our former ignorance of global change impossible to ignore—and, for many of us, inexcusable. The explosion of precise scientific measurements of numerous aspects of the earth's environment since the late nineteenth century, including such international baseline studies as that of the Millennium Project,[5] have provided ever-clearer pictures of how the earth is changing. Satellite measurements and photographs provide highly precise measurements and truly global perspectives. Meanwhile, news about even those things that happen in the remotest parts of the world can be accessed nearly anywhere through television, the Internet and mobile devices. Our village is global. We no longer have any excuse to claim ignorance of the changes humankind is imposing on the planet or of the consequences of those changes for our fellow human beings around the world and for all of life on earth. In the rest of this chapter, we look at some key elements of these changes and their causes, effects and prognoses.

POPULATION GROWTH

No one likes to talk about human population. Christians in particular are rightly wary of the ways in which population discussions have sometimes served to denigrate the value of human life and the blessing that children represent. Most of today's environmentalists also ignore the issue, because it is too controversial. Others understandably fear that talking about population in the context of environmental challenges serves only to mitigate our own responsibilities: it allows us to blame the world's problems on other people—or on too many people—rather than considering how our own lifestyles might need to change. The unpredictable nature of population growth and its effects (all too evident in how some past predictions have proven spectacularly wrong) and the assumption that it cannot be slowed except through coercive, top-down policies like those used in China have further contributed to the failure of the issue to find traction in contemporary discussions. Yet to ignore population growth is to ignore one essential part of the equation that at the very least has to be recognized as affecting life on earth, even if we were to conclude that there is nothing to be done about it or that other issues must take priority.

The world is undoubtedly getting more crowded. At the time Jesus walked the earth at the start of the first millennium there were an estimated 300 million people alive. Now, near the beginning of the third millennium, there are over 7 billion (7,000 million) people (see figure 2.1). What is perhaps most remarkable is how much of the growth in human population has happened very recently and at an exponential rate of increase: we did not reach 1 billion people until the turn of the nineteenth century, it took another 127 years until we added another billion, but recently we went from 6 billion to 7 billion in a mere twelve years. At the moment, the global population is increasing at a rate approaching a quarter of a million people every day.

Every additional person of course adds to the impact that we have on

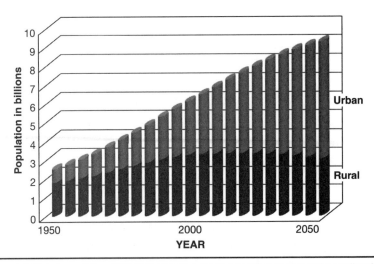

Figure 2.1. Global population 1950–2050 (data from United Nations). Note that the predicted 2.6 billion population increase from 2000 to 2050 is entirely in urban areas, with additional migration from existing rural to urban areas.

the planet. We need more food and more energy, we consume more natural resources and we produce more waste. Moreover, as Christians in particular must recognize, it is only right that those in the poorer parts of the world are enabled to gain improved living standards, better nutrition and better access to electricity, water and natural resources. All of this adds to the demands on the earth's resources. It drives home for us the need to manage the environment ever more carefully, to seek to live in ways that enable the flourishing of life on earth. We can no longer afford—if we ever could—to be careless, to assume that the ocean is so vast that fish stocks are endless and that it will absorb our waste products without change; that the soil can be made to produce however much we demand of it; that our aquifers will provide as much fresh water as we could ever want; or that our airborne pollutants will always be so dispersed as to have no effect.

The best estimates of population growth suggest that the rate of population increase has begun to slow a little. If things carry on as they are, it is likely that the number of people living on this planet will increase to just over 9 billion by the middle of this century (see figure 2.1)[6] and may exceed 10 billion by 2100. But the growth is very uneven. In high-income countries such as those of the European Union and North America, fertility rates have dropped to below replacement level. Population growth in those countries occurs only by immigration. The highest fertility rates are consistently in low-income countries such as those in sub-Saharan Africa, Yemen, Pakistan and Afghanistan.

The good news is that there are ways to address population growth that do not involve draconian governmental control or the denigration of families. The best way to stabilize population growth involves a three-fold strategy of improving healthcare, providing education and empowering women. Improved healthcare means that women are not required to have a large number of children in order to expect just two to survive (the stable replacement level is a little over two children per woman). Better education, particularly of women, who have traditionally lagged behind men in educational opportunities in developing countries, has also been shown to lead to reduced birth rates. And improved living standards generally, combined with the empowerment of women, tend to result in reduced family sizes. All of these factors are ones that Christians should be happy to sign on to and work toward. Indeed, Christian charities have historically been at the forefront of improving healthcare and education in low-income countries.[7]

In addition to the sheer increase in numbers, the other striking change in population demographics over the last two centuries has been a strong and increasing tendency for people to move from the countryside to urban conurbations or large cities, particularly following industrialization. In 2007 we passed the point where more people lived in cities than in rural areas (see figure 2.1). Over the four decades from 2011 to 2050, the world population is expected to increase by 2.4 billion, but those living in urban areas are projected to increase by 2.6 billion. So not only will cities absorb all the population growth, but there will be a continued move away from rural areas as well.

The benefits of cities—for individual opportunity, jobs, culture, creativity and (in well-managed cities) efficiency and even sustainability—are well known. One negative consequence of increasing urbanization, however, is that people are ever more divorced, not only physically but also mentally and emotionally, from the source of the food and water on which they depend for their very lives. As a result they are less likely to be mindful or understanding of the consequences to the natural environment of their actions and choices.

Increasing urbanization also means that the capacity for a single disaster such as an earthquake or flood to affect a huge and growing number of people is now vastly increased, even compared to a century ago. In 2011 there were 450 cities in the world with more than 1 million inhabitants, totaling 1.4 billion people. Of these about 60 percent, with a total of some 890 million inhabitants, are located in regions exposed to at least one risk of a natural disaster.[8] The number of people liable to be killed or displaced in each event is actually increasing year by year, despite our developing scientific understanding of such disasters. It is likely that within our lifetimes there will be a disaster that kills more than a million people; already we have exceeded two-thirds of that toll in a single earthquake in the mid-twentieth century when there were fewer huge cities.

LIMITS TO GROWTH?

In natural ecological systems, populations of species rise and fall dramatically with the supply of food. In his treatise *An Essay on the Principles of Population* (six editions between 1798 and 1826), the clergyman Thomas Malthus applied this principle to human beings. Malthus famously claimed that an increasing human population would inevitably outstrip agricultural production and lead to a return to subsistence farming; this came to be known as a Malthusian catastrophe. Malthus thought that a massive reduction in population was likely to be caused either by warfare as resources became scarcer, or by sickness or epidemics. If these did not sufficiently reduce the population, he wrote that "gigantic inevitable famine stalks in the rear, and with one mighty blow levels the population with the food of the world."[9]

Malthus made no specific predictions about when this might occur, but others following him have not hesitated to do so. For example, here is what Paul Ehrlich wrote in his 1968 book *The Population Bomb*: "The battle to feed all of humanity is over. In the 1970s and 1980s hundreds of millions of people will starve to death in spite of any crash programs embarked upon now. At this late date nothing can prevent a substantial increase in the world death rate."[10] Thankfully, Ehrlich was wrong. Although it is true that hundreds of millions of people have died of hunger, malnutrition and related illnesses in subsequent decades, this is not the result of a "Malthusian catastrophe." Rather it is due to a complex mix of local disasters, corruption, injustice and other factors that mean food is not distributed equally around the world. In fact, the increase in food production brought about by the "Green Revolution" meant that there was always sufficient food produced globally, but its distribution was uneven and inequitable.

It is important to observe, however, that the term "Green Revolution" is rather a misnomer. The Green Revolution resulted primarily from the massive use of fertilizers, pesticides and fossil fuels. All of these carry deleterious effects that are now beginning to be felt acutely (see the sections below on nitrogen and land use, and chapter 3). Most of these resources are not renewable and hence their use is not sustainable over the longer term. Much hope is now invested in genetically engineered crops as a potential means of providing increasing yields in the future, although this promise has yet to be realized. In any case, ongoing use of genetically identical crops over wide areas (monocultures) carries with it the increased risk of catastrophic failure from such things as unforeseen disease.

Nevertheless, food scarcity is not presently a trigger for huge population decreases (at least globally), although the threat of pandemics remains a real and present danger for humankind. More generally, opinions differ as to whether Malthusian ideas can even be applied to human populations, since the counterargument is that humans are endlessly creative and so are always likely to find technological ways of avoiding disaster. That, for example, is the view of Mark Lynas in his 2010 book *The God Species*, which he subtitles *How the Planet Can Survive the Age of Humans*.[11] However, as Jared Diamond has shown in his book *Collapse: How Societies Choose to Fail or Succeed*, there is evidence historically that populations and civilizations have indeed died out because of overuse or misuse of environmental resources in the midst of other social and political failures.[12] At the very least, the highly complex interactions between the different elements of the biosphere mean that any simple technological solution for one problem can easily exacerbate another. It would be foolish to respond to the false prognostications of those like Ehrlich with a naive optimism in the ability of our own species or civilization always to evade the challenges that have faced all others.

Although a few technocrats believe that endlessly increasing populations will inevitably bring with them the ingenuity to find new solutions to global food supply, most people believe that there must be a sustainable limit to the population on earth if all are to live with dignity and sufficient resources. Estimates of the sustainable carrying capacity of the earth vary widely according to what assumptions are made, but most lie in the range of 6 to 14 billion people, with the most common being 6 to 8 billion, roughly the same as the present global population.[13] The question that all such estimates raise, however, is just what we mean by *sustainable* and to what extent we include humankind's impact on the rest of life on earth. It is to this latter issue that we now turn.

Assessing Humankind's Impact on the Earth

Scientists have recognized for some time that humankind is playing an increasingly powerful role in shaping the processes that affect all life on earth. But the staggering scale of our influence has only recently become apparent. Humans now exert by far the biggest biological and geological influence on the earth. Our reach is all-pervasive. Among us we now move twice as much earth and rock per year as all the natural erosion processes of the earth put together. We move every year on average six tons of earth material for every person on this planet; in the United States, the average is up to thirty tons per person. Over the past five thousand years the amount of earth moved by humans would make a massive artificial triangular mountain as high as Everest, the tallest mountain on earth, with a base over 70 miles (110 kilometers) long on each side. We are changing the global climate at a rate never before seen in the geological history of our planet. And our species alone is using between one quarter and one third of the entire planet's primary productivity of plants (the way the sun's energy is converted to food, on which the whole food chain depends).

The scale of these effects means that it has become vital for us to be able to assess accurately the impact of our activities on the earth if we are to make wise decisions about our future. One attempt to quantify and collate much of the relevant data, involving thousands of scientists from all over the world, was the Millennium Ecosystem Assessment.[14] Its objective was to quantify the effects of ecosystem changes on human well-being. Another example, to which we will return in the next chapter and which again involves thousands of contributors from around the globe, is the regular series of authoritative reports produced by the Intergovernmental Panel on Climate Change (IPCC).[15] These reports, by scientists from a wide range of social, cultural, political and religious backgrounds, are among the most striking examples ever recorded of global scientific consensus on the common problems facing humankind.

Baseline reports like these of where we stand now are hugely important. But, as we will see, it is often difficult to gain a consensus on what to do about the dangers such reports highlight, and it is all the more difficult actually to bring about the changes that are found to be necessary. How can individuals be encouraged to change ingrained habits, let alone whole nations or even the entire world? It is easy to lose heart in the light of such statistics as those reported in the United Nation's 2010 *Global Biodiversity Outlook*, which summarizes the (lack of) progress made by parties to the international Convention on Biological Diversity (CBD). It makes for grim reading: despite some local successes, *none* of the twenty-one subsidiary targets to the CBD's 2002 goal of achieving a "significant reduction" in the rate of biodiversity loss by 2010 had been met. Worse, ten of fifteen indicators tracking biodiversity show negative trends.

It is worth remembering the occasional international success stories too, however, as they are easily forgotten in the face of so many new challenges. For example, following scientific consensus that the stratospheric ozone layer that protects the earth from ultraviolet radiation was growing dangerously thin as the result of widespread use of chlorofluorocarbons (CFCs), all UN-recognized nations signed up to the 1987 Montreal Protocol, which phased out CFCs worldwide.

There are undeniably many more failures than successes, but this does not lessen our responsibility to consider carefully the present state of our planet and to reflect on the possible implications for ourselves and our societies. The remainder of this chapter, then, provides some details about the state of the planet, drawn largely from some of the international scientific assessments mentioned above. We also consider along the way whether any consensus is developing about how we might address the challenges identified in these reports.

Several related approaches have been developed for quantifying and tracking the impact of humans on the earth. One popular method is to attempt to calculate the "footprint" of human activities and to devise ways to keep this within sustainable bounds.[16] Another is to look for "tipping points" beyond which the earth system will run away no matter what we do, so that once such tipping points are identified we can attempt to avoid them.[17] A third approach (mentioned in chapter 1) is to map out boundaries to our various activities, such as land use or production of greenhouse gases, within which we can continue to live safely, but beyond which severely deleterious consequences are likely to develop.[18]

Inevitably there are dangers associated with setting such boundaries. For one thing, they are largely arbitrary choices. Who is to say that the loss of ten species per million every year is acceptable, but more than that is not? For another, the interactions among different parts of the ecosphere are so complex that not only may we fail to recognize or understand those interactions, but our choice of issues on which to focus may be quite wrong. We may miss entirely the elephant in the room. Furthermore, while thresholds are comforting for policy makers, they can also be risky. Waiting for thresholds to be crossed in the mistaken notion that there is no problem if we do nothing may merely allow the continuation of bad practices that might have been prevented. Also, defining boundaries on a global scale necessarily ignores the impact of local circumstances in exacerbating or ameliorating the problem of managing scarce resources.

Nonetheless, measures of our collective impacts on the earth and their likely effects, trajectories and potential boundaries still remain helpful and perhaps necessary if we are to understand and seek to address issues that are genuinely global in scope. We must, of course, acknowledge their shortcomings and recognize the need to join global perspectives with local ones. Meeting the challenges of our century will require citizens who are internationally aware and involved while at the same time committed to their own bioregions and local communities. The focus in the following sections, however, is necessarily on a few of the major global issues that face humankind. And for that task, we will occasionally draw on the work of those advocating "planetary boundaries" as one helpful if imperfect way of summarizing and quantifying the relevant data.

BIODIVERSITY

The loss of furry mammals captures the public eye. The extinction of another few dozen beetle species—quite possibly not yet even known or cataloged—may not. Yet such losses may be equally troubling from an ecological perspective—and from a Christian perspective too, if we consider the diversity of creation as having value before the God who cares for every lily of the field, every sparrow that falls. Human-caused extinctions are nothing new, of course: whenever humans have colonized new areas, we have inevitably had dramatic effects on natural ecosystems. In North America early humans wiped out, among other things, mammoths, several species of deer, moose and antelope, all ten species of North American horses and nearly all of the vast herds of bison that once roamed the plains. In New Zealand the colonizing Polynesian people who arrived seven hundred years ago destroyed much of the unique island ecosystem, which was then dominated by various unique species of bird, including the giant flightless moas and eagles with record-holding ten-foot wingspans. Within a century all of these New Zealand birds, together with half the other terrestrial vertebrates on the islands, were dead.

As a result of past extinctions, our children will never experience the beauty of a golden toad (not seen since 1989, now presumed extinct), the quirkiness of a dodo (extinct since the late seventeenth century), or the spectacle of a vast cloud of migrating passenger pigeons up to 2 billion strong darkening the sun and taking hours to pass overhead (as reported from North America in the nineteenth century, before the birds were slaughtered wholesale just so that their carcasses could be used for fertilizer). From being one of the most abundant birds on earth in the nineteenth century, passenger pigeons were hunted to extinction. Martha, the last known survivor of her species, died in captivity on September 1, 1914, in Cincinnati, Ohio.

While there may be nothing new about human-caused extinction, the rate at which species are now going extinct due to the effects of human beings is truly unprecedented. Compared with the earth's normal "background" rate of extinction, the current rate is estimated to be anywhere from one hundred to one thousand times higher as the result of human activity.[19] In total, there are today one-third fewer wild animals on the planet than there were even just forty years ago.[20] Most of the extinctions taking place are the result of habitat degradation and land use changes (see more on this below under "land use"); for example, half of the earth's forests that existed after the end of the last ice age nine thousand years ago are now gone. The ongoing destruction of tropical forests alone—the most biologically diverse places on earth—is in fact the biggest contributor to today's unprecedented rate of species loss.

On average one species on earth goes extinct every eight hours. Up to 30 percent of all mammal, bird and amphibian species are threatened with extinction this century. We are rapidly heading toward (or, according to some biologists, are in the midst of) a mass extinction similar in magnitude and scope to that which killed off the dinosaurs 65 million years ago. In that case, an asteroid was the most likely culprit; this time, it is us.[21]

Not all is doom and gloom. Around the world conservationists are making strenuous efforts to prevent extinctions, and there are some heartwarming tales. To take just one example, in 1980 there were only five surviving black robins in the world, living on Little Mangere Island off New Zealand, and among them only one fertile female, "Old Blue." A dedicated team of people protected the surviving birds round the clock. Eventually they took the first clutch of eggs each year and hatched them using tomtits as surrogate parents while the robins laid a second clutch.[22] Black robins, though still endangered, came back from the brink and now comprise a population of 250.

But let us return to the question implicitly raised at the beginning of this section: Why should we care about loss of biodiversity? Apart from aesthetic or moral considerations, one reason is that as humans we rely on living systems to keep our air breathable, our water drinkable, and to provide us with sufficient food. Loss of biodiversity makes the ecosystems vulnerable to diseases and other disasters that could wipe out species on which we depend. The great American conservationist Aldo Leopold reflected on the implication of this when he compared wildlife management to "intelligent tinkering" in which the first rule is to save all the parts—"every cog and wheel."[23] Loss of even a few species in a complex interacting ecosystem can greatly reduce its resilience to change and make it vulnerable to catastrophic and irreversible failure. To cite just one example, there are concerns that the pressure on many species of bees could lead to significant social and economic losses.[24] Bees pollinate 75 percent of plants, including cash crops in Britain with a value of over £500 million. Yet bees are under pressure from such factors as monoculture, use of pesticides, habitat degradation, diseases and climate change. Already, wild honeybees are thought to be nearly extinct throughout the British Isles, and the number of managed honeybee colonies in the UK fell by 53 percent between 1985 and 2005.[25]

From a purely utilitarian perspective, the wide range of plants and animals provide resources that have huge economic value, which at present we take for granted. In what is perhaps a rather silly but nonetheless widely influential calculation, the total value of the biosphere in 1997 was estimated at $33 trillion per year, roughly double the world's total Gross National Product (GNP) at the time.[26] (Of course, without the biosphere there would be no GNP at all, nor any human beings to assign monetary values to those things that sustain life itself!) But to take a more specific example, the economic benefits that are lost as a result of the world's fisheries being managed unsustainably are put at $50 billion each year. And for every species that we drive to extinction, we deprive future generations of their potential value. US president Teddy Roosevelt expressed the utilitarian motivation well in 1907 when he claimed, "To waste, to destroy our natural resources, to skin and exhaust the land instead of using it so as to increase its usefulness, will result in undermining in the days of our children the very prosperity which we ought by right to hand down to them amplified and developed."[27]

From a Christian perspective there is a more compelling reason than the strictly utilitarian one for seeking to prevent the unnecessary loss of other species: we have been set the task as God's image bearers to care for his creation, to be vicegerents in its governance on his behalf. The Bible is suffused with images of God caring for his creation with all its complex interactions. Nor does God do so just for the sake of humankind. God "water[s] a land where no one lives, an uninhabited desert, to satisfy a desolate wasteland and make it sprout with grass" (Job 38:26-27). According to Psalm 147:9, God provides food not just for cattle but "for the young ravens when they call." He is also quite happy with the food chain he has ordained on earth, including even those carnivores that most seem to threaten human beings: Psalm 104:21 tells us that when lions roar, they seek their food from God.

Economic valuations have their place, and utilitarian approaches can be useful in broadening support for the sorts of policies necessary to preserve biodiversity. Yet, for Christians, to reduce all of the beauty and fruitfulness of God's creation only to a monetary value is ultimately an affront to the Creator and Sustainer of all that exists. It is perhaps symptomatic of a generation that has turned its back on God and chosen instead to put its trust in material things while forgetting or ignoring the Creator and Provider of those things. Aldo Leopold once observed that one of the penalties of an ecological education was "to live alone in a world of wounds."[28] An ecological education makes a student aware of things that most people never notice; it becomes impossible to ignore what is lost when land is torn up, waters polluted and species after species disappear, never to return. But even for those of us who lack a formal education in biology or ecology, Christians of all people ought to be able to join alongside biologists like E. O. Wilson who act as advocates on behalf of the diversity of nonhuman life—including not just charismatic large mammals but beetles too.[29] If, as this book explores, such life has value before God, we actually have all the more reason to mourn its loss and to do all we can by God's grace to nurture and to keep what is left.

WATER

Water is an essential requirement for any form of life. The earth is blessed with huge volumes of water at the surface, which is one of the main reasons why it is habitable. Seventy percent of the earth's surface is covered by sea, and the weather continually recycles fresh water from rain through rivers and soil. One of the remarkable features of our planet is that throughout the four-billion-year history of life, its surface temperature has stayed between the limits of 0°C (32°F), when everything would freeze, and 100°C (212°F), when all the water would boil off into space. This is despite the intensity of the sun having increased by one-third over the same period. Had the temperature strayed outside the range 0-100°C (32-212°F), the earth would have ended up sterile with no life as we know it. The primary reason for this stability is feedback from myriad biological processes. From a Christian perspective, it can be seen as God's providence in sustaining a world in which life can flourish.

But even so, the water supply is a finite resource. We are polluting the seas with rubbish and chemicals and allowing their acidity to change dramatically as a result of greenhouse gas emissions. So much water is extracted from rivers that already a quarter of the world's river basins run dry before they reach the sea. The Aral Sea in the former Soviet Union, once the world's fourth largest lake, is now reduced to a quarter of its former size and is desolated due to diversion of water from the major rivers flowing into it. Once the Aral Sea was a major fishery yielding 44,000 tons of fish per year; now it produces none.

Humans currently use nearly 1,000 cubic miles (4,000 cubic kilometers) of fresh water globally every year for drinking and agriculture. Two-thirds of this water is used for agriculture. The predicted population increase and associated increase in urbanization will place increasing pressure on water supplies as more food will be required to feed the extra people.[30] In this case, the global statistics alone can actually mask what is a greater tragedy at local, human scales. By 2030 it is estimated that nearly half the world's population will be living in areas of high water stress.[31] Today only half the world's population has a supply of clean fresh water on tap. The other half has to carry water from distant and sometimes polluted sources. Today 1.8 million people every year die from diarrheal diseases, with 90 percent of them being children under five years old. Most of them die because they have unsafe water supply or inadequate sanitation. That is more than three people dying every minute of every day from lack of something most of us reading this book simply take for granted.

In the oceans, our garbage now extends to their farthest reaches: floating plastic bags and other human detritus are a common sight even in the middle of the Pacific, far from land.[32] The greatest threat to the oceans, however, is invisible and much worse. Their acidity is increasing inexorably, largely due to growing amounts of carbon dioxide being absorbed in them as a result of our burning of huge quantities of fossil fuels. The oceans absorb approximately one-third of the extra carbon dioxide produced by humans. Since pre-industrial times this has caused the acidity of the oceans to increase by about 30 percent, and this increase is accelerating.[33]

Why should we worry about increases in ocean acidity? The main reason is that acidification will directly affect a wide range of marine organisms that build shells from calcium carbonate. These include tiny plankton, mollusks and corals that range from the shallow to the deep sea. These organisms are the basis of the food chains that are fundamental to the overall structure and function of marine ecosystems. Any significant changes could have far-reaching consequences not only for the ocean but for millions of people who depend on its food and other resources for their livelihoods. There are significant unknowns; we simply do not know what the long-term effects of increasing acidification of the oceans will be on the ecosystems. But the current indications are not good, with significant damage already occurring to coral reefs, which harbor the richest marine biota.

Nitrogen

It might seem strange to single out an inert gas, nitrogen, as a critical component of the biosphere and of the future of humanity, especially since there is apparently such a lot of it around—78 percent of the air we breathe is composed of nitrogen. But in fact, for most of human history, the lack of sufficient nitrogen in the soil has been the main limiting factor for plant growth. It was only in the twentieth century that this limitation could be overcome by the large-scale production and application of industrial crop fertilizers. A chemical method known as the Haber-Bosch cycle was developed to extract nitrogen from the atmosphere, a discovery that won Fritz Haber the Nobel Prize in 1918.

Largely as a result of the availability of synthetic fertilizers, combined with the use of fossil fuels to power machinery and the development of high-yielding hybrid crops, the "Green Revolution" of the mid-twentieth century led to a huge increase in the amount of food the world could produce. Between 1950 and 1984, global grain production increased by more than 2.6 times.[34] As we already observed, it was this Green Revolution that staved off a chronic shortage of food that would otherwise have resulted from the fourfold increase in population over the same period. But it carried many negative environmental effects, including soil erosion, pollution of ground and surface water, and public health problems caused by pesticides.

Synthetic fertilizers and the burning of fossil fuels now add 160 million tons of nitrogen to the environment every year. That is more than the global natural biological fixation of nitrogen on land (110 million tons) or in the ocean (140 million tons). Humans are responsible for doubling the turnover rates of the nitrogen cycle of the entire earth.[35]

The benefits of this human production of nitrogen in greening the earth and enhancing crop production are clear, but there are also considerable downsides with uncertain but potentially dangerous consequences. Much of the reactive nitrogen eventually ends up in the environment, where it pollutes waterways and coastal zones, adds to the local and global pollution burden in the atmosphere and accumulates in the biosphere. The main immediate results include global acidification and stratospheric ozone loss, the generation of algal blooms when fertilizers are washed off fields into watercourses resulting in oxygen depletion of rivers (called eutrophication), and a host of unpleasant side effects such as production of photochemical smogs and ecosystem acidification. The localized effects of our use of nitrogen fertilizers are often obvious. Residents living alongside Lake Spokane near Jonathan Moo's home in Washington, for example, often complain about huge algal blooms that occasionally become dangerously toxic; but such is the inevitable result when all of us who live in the river's watershed rely on nitrogen and phosphorus fertilizers to grow crops more intensively or (less defensibly, perhaps) to keep our lawns looking green.

The only effective method of removing reactive nitrogen is to combine it with another nitrogen atom to produce the inert gas N_2. That occurs naturally by certain microbes, most of which live in low-oxygen aquatic environments. Wetlands are a good example of where such microbes flourish, so there is a strong case to be made for preserving wetlands, flood meadows and deltaic regions simply because of their nitrogen-fixing abilities. As always, many environmental issues are tightly interlinked. With our penchant for channelizing waterways, draining bogs and concreting over everything, we often do not realize the collateral damage we may be causing.

Reducing the volume of nitrogen that we produce requires, of course, that we reduce the amount we use. This need not mean entirely giving up the huge advantages of using fertilizers. Just as insulating your house and fitting double glazing both reduces your heating costs and reduces the amount of carbon dioxide that we put into the atmosphere, so reducing the amount of fertilizer farmers need to use is a win-win result. It can be done by breeding plants that take up nitrogen more selectively (either by conventional breeding or by genetic engineering) and by adopting methods of irrigation that reduce the runoff of water and fertilizer. This has the obvious added benefit of helping us use less water. At present two-thirds of the nitrogen applied in fertilizers is almost immediately washed away, so there are huge gains to be made in this area.

The Planetary Boundaries group considers that nitrogen use is one of only three of the ten critical parameters they identify that have already exceeded a dangerous boundary (the others are biodiversity loss and climate change). In their view we have surpassed the safe limit by a factor of about four, although as they say it is hard to know the precise results of the huge increase we have already caused in nitrogen fluxes. Whatever else we decide to do, reducing our nitrogen use is an important objective and one that could be met given a willingness and commitment to do so.

Food

Eating is, of course, one of our most basic needs. It is also among the most important social and cultural activities of society. It is no surprise, perhaps, that shared meals play such an important role in Scripture.[36] Several of Jesus' miracles involve food or wine—and they are in contexts where the miraculous provision is not a matter of life and death, but simply one of enabling people to enjoy the blessings that God intends humans to have. A shared meal is at the heart of the commemoration of the Passover, celebrating God's rescue of his people Israel from slavery in Egypt and the beginning of their journey to the Promised Land. Jesus gave new depth of meaning to the story when he instituted the sharing of bread and wine that the Christian church celebrates in the Eucharist as the chief way for us to remember together his incarnation and sacrifice for us on the cross.

The very basic activities of eating and drinking tie us unequivocally to the land. As many recent writers have lamented, it is one of the travesties of modern urban life that so many of us are disconnected from contact with the very means of our own survival. It is perhaps one of the reasons why many of us know so little about what we are doing to the earth in the process of getting our food supply. That disconnect is also perhaps why modern city-dwellers are often careless of giving heartfelt thanks and praise to God for a successful harvest, or for the simple provision of a square meal.

As any biologist knows, the availability of food supplies is often the main limiting factor in population growth in natural ecological systems. Humans have taken that process of food production into their own hands, and this is arguably the most important factor in our ascendancy to domination of the biosphere. The first big step came in the development of cooking, which enabled humans to get far more energy out of food than eating it raw. Then came animal husbandry and agriculture, which allowed humans to settle in villages, towns and cities as they took charge of the production of food in the surrounding region. Selective breeding, both of plants and animals, allowed a steady increase in population. Then, most recently, the ability to produce inorganic fertilizers and to capture the energy from burning fossil fuels led to the Green Revolution with its huge increase in crop production.

The amount of energy produced by an engine burning one gallon of gasoline produces as much work as about a month's labor of one person. Mechanization powered by fossil fuels is an astonishingly powerful tool. In terms of the energy we each utilize in our daily lives, either personally or by proxy in the foods we eat and resources we consume, people in high-income countries like the USA and UK are relying on the work of the equivalent of two hundred human laborers. The consequence of our dependence on machinery in modern agriculture is that so much fossil fuel is used in food production, packaging and distribution that it is estimated that seven to ten kilocalories of energy are expended for every kilocalorie of food consumed in the United States. We are, as one recent writer has said, essentially eating fossil fuels.[37] This cannot continue indefinitely: we are close to, or past, the peak in conventional oil supplies, water supplies are becoming increasingly stressed, nitrogen levels from artificial fertilizers are already dangerously high and, as we will explore in the next chapter, continued global warming will eventually lead to dramatic decreases in crop productivity globally.

Despite these challenges, however, improving the efficiency with which we grow and distribute food is not an impossible goal—far from it. An astonishing one-third of all the primary crops grown in fields is lost or wasted and not consumed, amounting to 1.3 billion tons of food each year worldwide.[38] In low-income countries, most is lost in harvesting and in storage (to pests or rot). In high-income countries, most is wasted by retailers or consumers who throw out surplus food before it is eaten, or allow it to rot. We will need to become more careful stewards of the earth's resources, to develop new crop strains and methods of irrigation, fertilization and pest control that are sustainable in the long term. But we are unlikely to be willing to pay the necessary extra costs and to invest in sustainable agriculture unless and until we acknowledge the scale of the problem to be addressed.

In high-income countries the cost of food in real terms is at an all-time low. Today luxury food is literally a throwaway commodity in high-income countries. A millennium ago, most of the working lives of the great majority of people were spent in simply providing sufficient food and shelter for their families. As recently as 1930, the average American spent 24 percent of their disposable income on food; now it is only 9 percent. Food is a little more expensive in Europe, but not much; Britons, for example, only spend an average of 11 percent of their income on food. As you move to poorer countries, however, the percentage predictably goes up—to 27 percent in South Africa, over 50 percent in India and over 70 percent in Tanzania.[39] Globally, more than half the world's population spends 50 percent of their income on food. In those circumstances, the upward pressure on food prices from such things as the production of biofuels or a poor grain harvest due to climate change is more than just an inconvenience.

The other aspect of food that carries with it strong ethical and moral implications is the fact that today about 800 million people in the world are malnourished. Never a year goes by without reports of another famine, often in Africa, with all of the attendant pain, anguish, sickness and death it causes. Yet this is not—at the moment—for lack of a global supply of food. In addition to the losses and waste mentioned above, there are an estimated 1 billion people in the world who are obese and whose health and life expectancy suffer from being overweight. The causes of this inequality in food distribution are complex; the mix of factors usually identified include the effects of globalization, political unrest, greed, domination of markets by large international corporations with profit as often their sole driving force, and protectionist trade policies that in some cases lead to mountains of food being destroyed just to keep prices up. But, as always, it is the poor and disenfranchised who suffer most.

Some of the trends that lead to food scarcity are readily identifiable. One, for example, is that crops are increasingly being grown for biofuels. In the United States, more than 40 percent of the corn crop in recent years has gone into producing ethanol, which can be mixed with petroleum to fuel cars. The rationale for this is said to be to reduce the impact of burning fossil fuels and so help reduce climate change. But it is not clear that the use of corn ethanol leads to any net reduction in the burning of fossil fuels. What has become clear is that using cropland to grow fuel for our cars leads to large price increases that particularly affect the poor, creating food shortages and even food riots in many countries. The amount of grain needed to fill one tank of a typical sport utility vehicle with ethanol could feed a person for a year.[40] The global rush for biofuels also involves tearing down huge tracts of tropical rainforest, as has happened in Indonesia and Malaysia to make plantations for oil palm (one-third of which is used for biofuels, the rest mainly for processed foods). It is a sobering reminder of the danger of seeking easy solutions for one issue without considering the wider implications and consequences.

Last, we cannot leave the topic of food without considering the increasing consumption of meat and dairy products. As people get more affluent they tend to eat more meat. Today there are more than 1.4 billion head of cattle in the world, with a population of just over 7 billion people. Only 3 percent of terrestrial vertebrate flesh is wild. Of the rest, one third is human and two thirds are domesticated animals. To produce one kilogram of meat takes nearly ten kilograms of vegetable protein, and therefore it also requires more than ten times the quantity of water that would be used to grow an equivalent amount of vegetable protein. In the second half of the twentieth century, about 40 percent of the world's grain production and 75 percent of its soy production was fed to livestock, and these percentages are set to rise still further as demand for meat grows much faster than population growth.[41] Given the pressures on both water and fertilizers that we have already discussed, the world can scarcely afford such profligacy.

Livestock also has a direct effect on global climate change. It is estimated that 18 percent of human-induced global climate forcing is produced by greenhouse gas emissions generated by livestock, if the entire cycle from farm to market is taken into account.[42] As we discuss in the next chapter, that is more than the greenhouse effect of all the forms of transport in the world added together—all the cars, trucks, buses, trains, ships and airplanes. One-third of this impact on climate is caused by the release of carbon dioxide into the atmosphere as a consequence of the massive deforestation that occurs to produce grazing land and croplands for growing feedstuffs. A further quarter of the climate impact is generated by the methane produced by livestock as animals digest vegetation, while another quarter is ascribed to the breakdown of manure, which releases both methane and nitrous oxide. Ironically enough, the huge increase of meat consumption is not only bad for the environment, but bad for the people eating large quantities of meat too. The incidence of heart attacks and strokes increases dramatically in those who eat a lot of red meat.

The purpose of this discussion of livestock is not to suggest that we must all become vegetarians, though some do in fact choose to do just that. Others commit to eating less beef, and only from rangeland, grass-fed cattle, which, if managed well, have far less damaging impacts (and in some places may actually be better for the land and wildlife than growing grain). The main point of this discussion, however, is to remind especially those many of us who live far removed from the sources of our food that everything we do (and eat) has consequences, both for the environment and for other people. Moderation in the amount of meat we eat, considering the sources of that meat and switching to other meats such as fish, chicken and ostrich can go a long way toward reducing the deleterious effects of our eating, without reducing the enjoyment there is in sharing a good meal.

LAND USE

All of the factors discussed in this chapter—biodiversity, water, nitrogen and food, each of critical importance to the future of humankind and life on this planet—relate in one way or another to how we use the land. Nearly 85 percent of the global ice-free land surface has now been directly influenced by humans, and some biologists—interpreting "human influence" more broadly—claim the figure is nearer 100 percent. Our influence can be benign as well as harmful, of course, but in any case it is impossible to miss the sheer scale of our impact on the earth. For example, currently we use about 30 percent of the entire global ice-free land surface to raise cattle.[43] A further 12 percent has been converted to cropland. The Planetary Boundaries group estimates that the safe upper limit of cropland with intensive agriculture is about 15 percent of the global availability, so we are fast approaching that boundary. However, most of the consequences of land misuse are felt at local and regional scales rather than global scales, so the suggested global boundary of 15 percent can mask what are often locally devastating consequences.

People live on land. The extent of nations is defined by their land boundaries, and wars are fought over them. If the patch of land on which you live and from which you (or your region) derive your livelihood is made uninhabitable, the consequences may be catastrophic. One change that may make it impossible to live in a particular place is a change in climate. This might bring either long-lasting droughts or devastating flooding, both of which can, and do, lead to regional famines. Another change might be rising sea levels, which can cause the land to be flooded. Flooding by storm seas can deposit so much salt on the land that it can no longer be used for agriculture. In Bangladesh, 30 million people live in a deltaic region within one meter of sea level. It is impossible to protect such a delta from storm seas by building a barrier, so flooding becomes inevitable. And some people living on low-lying islands such as the Maldives risk losing their entire islands to rising sea levels.

Apart from the effects of a changing climate, which are treated more fully in the next chapter, one of the greatest threats to agriculture and to the health of the land generally is irreversible soil loss. In the latter half of the twentieth century nearly one-third of the arable lands worldwide were so heavily affected by water erosion (by far the biggest proximate cause of soil loss), wind erosion or chemical and physical degradation that they had to be taken out of cultivation. The primary causes of this frightening loss of topsoil around the world are unfortunately linked to the same things that made the Green Revolution and intensive agriculture possible. Overall 8 million square miles of land worldwide have been degraded as a result of unsustainable agricultural practices; that is nearly one hundred times the area of Britain, or the area of the whole of China and the United States combined. The degradation of two-thirds of this area is estimated as moderate or worse.[44] What is needed now is a revolution in agriculture that is genuinely "green," that is good for the land in the long term and is able to go on feeding the world's growing population.

It was the misuse of semi-arid prairie grassland by deep plowing for crop planting that removed the roots of the grass that retained moisture during droughts and winters. This led to the severe soil erosion and massive dust storms in America and Canada during the 1930s. The resultant Dust Bowl caused a mass exodus of 2.5 million people from the Plains states by 1940. Today, in the very different ecosystem of rainforests, it is feared that if too much of the Amazon is felled, the region could transform irreversibly from its rich diversity of life to semi-arid savannah. Just in the years since 1970, about 20 percent of the Brazilian Amazon alone has been cleared, representing the loss of 280,000 square miles of forest[45]—an area bigger than Texas or nearly three times the size of the entire United Kingdom. Most of this forest is cut or burned down to make room for cattle ranching; some of it is used for agriculture and lumber. As we saw earlier, the destruction of tropical forests generally is a major contributor to the dramatic decline in our world's biodiversity. And as pressures increase to grow more food (and biofuels), ever more forest is cut and ever more marginal land is brought into cultivation, heightening the danger of irreversible degradation in areas that have little chance of recovering.

What happens when crops extend across ever larger areas? Perhaps the most worrying aspect is that by destroying natural habitats intensive farming can threaten the ability of varied ecosystems to self-regulate and maintain the living biosphere. Natural landscapes filter water, produce oxygen and sequester carbon (important in an era of increasing emissions of carbon dioxide), as well as provide the habitats—the homes—for the rest of life on earth.

Meanwhile, the ever-increasing development required for our growing population further contributes to the reduction of natural habitats and disrupts the hydrology and ecology of all those areas covered with concrete or asphalt. Development and the attendant building of highways has cut apart previously connected regions, preventing the free movement of animals. The resulting disconnected, shrinking patches of natural habitat leave plants and wildlife isolated and all the more vulnerable to the threats of a changing world.

Unsustainable practices of agriculture and development are the major causes of land degradation today, but there are other factors too. History reminds us of the potential consequences, for example, of unrestrained and poorly managed resource extraction. Oman has rich copper reserves near the surface that were mined and smelted by the Persians and their successors using wood from the heavily forested country. But in the tenth century A.D. smelting came to an abrupt halt. Why? It was not for a lack of copper, it seems, but largely because the whole country had been deforested.

The long-lasting, damaging effects of poorly managed mining and resource extraction continue to be felt in many parts of the world, and not just in developing countries. Each spring, tundra swans pause in their northward migration to feed in marshes alongside the Coeur d'Alene River in North Idaho not far from Spokane; but within days, this otherwise beautiful and apparently pristine landscape is littered with their dead bodies. The culprit? Heavy metals—millions of tons of mining waste from local silver mines—have poisoned the area's marshes. As the swans feed on plant roots and tubers, lead accumulates in their bodies until it shuts down their digestive systems; eventually they die of starvation (see figure 2.2). Downstream and across the state line, meanwhile, heavy metals from the same mines and PCBs (polychlorinated biphenyl) from ongoing industrial and wastewater pollution mean that children playing on beaches along the Spokane River and fishermen tempted to eat the river's trout face the risk of cancer, liver disease, neurological and behavioral complications, and reproductive and endocrine system problems.

It might be hoped that nowadays the governments of "developed" countries at least know better, that we have learned our lessons and that the long-term impacts of our use of the land are now always considered carefully and weighed against the potential benefits. Yet whenever the economic incentives are strong enough, it seems we make the same mistakes as those before us. To take but one recent example: whatever might be the merits and drawbacks of hydraulic fracturing (fracking) to extract natural gas from underground shale formations (a process in which a high-pressure stream of water and chemicals, many of them toxic, is used to split open rocks containing natural gas), the need for new sources of energy and the potential for quick profits has meant that the industry is expanding by leaps and bounds with little regulatory oversight. There is, as yet, little reliable scientific analysis of the potential impacts of fracking on the environment, including on the aquifers that supply drinking water to millions of people. In the United States, the Energy Policy Act of 2005 in fact specifically exempted hydraulic fracturing from the Safe Water Drinking Act, effectively preventing the Environmental Protection Agency from using what it calls its "central authority to protect drinking water."[46] As authors of a recent study put it, the lack of adequate study and oversight has made the

Figure 2.2. Dead tundra swan on the Lower Coeur d'Alene River, Idaho, poisoned by lead polution during its migration stopover. (Kathy Plonka/*The Spokesman-Review*)

worldwide increase in hydraulic fracturing "an uncontrolled health experiment on an enormous scale."[47]

It is sobering to consider the scale of humankind's effects on the surface of the earth, and this section has necessarily focused on the very real threats and challenges that we face as the result of our carelessness and ongoing overuse and misuse of resources. All need not be gloom, however. It remains quite possible, for example, to farm the land in ways that preserve its soil and nutrients, to develop crops that are suitable for different environments and to end the policies that prevent sustainable agriculture from being practiced more widely. While the tropics continue to face deforestation on an epic scale and many northern pine forests are being plagued by fatal insect outbreaks horrifying in their extent, in other parts of the northern hemisphere—the northeastern United States, for example—new forests have reclaimed much land formerly cleared. The growing discipline of restoration ecology is finding ways to help scarred land to heal and degraded habitats to be restored or created anew. Species once on the verge of extinction have recovered. To take a high-profile example, the once-endangered gray wolf has come back in sufficient numbers in parts of the American West that it is again being legally hunted and trapped in places like Idaho and Montana. In Yellowstone National Park, the reintroduction of wolves after a seventy-year absence has meant the beginning of a rebalancing of an ecosystem, the restructuring of elk populations and an accompanying recovery of vegetation that had been overbrowsed for decades.[48]

Even as we rightly worry about the ongoing destruction and fragmentation of natural habitats, over a tenth of the earth's land area has already been designated as a protective area of one kind or another (although the level of protection varies considerably, and the oceans lag far behind with only about 1 percent protected).[49] Astonishingly, the world's remaining biodiversity hotspots cover only a small area of the earth's surface. Nearly half of all known species of vascular plants and over one-third of all mammal, bird, reptile and amphibian species are found in just 1.4 percent of the earth's surface.[50] With enough political will and international cooperation, it ought to be relatively straightforward to conserve such a small area with such a high reward in biodiversity.

We will not have the will to take such steps, however, unless and until we have faced squarely and honestly the reality of our growing impact on life on earth. So we encourage readers to go on reading beyond the brief overviews we have given in this chapter, to follow sources given in the endnotes, and to keep up with at least some of the sorts of studies we have referenced here that pique your interest and concern. Many of us manage in our busy lives to track closely our favorite sports teams; we are kept constantly up to date on the doings of celebrities; and we follow the gyrations of financial markets with fascination and a mixture of hope and dread. Yet mainstream media leaves us frighteningly ill-informed about the things that are most vital to life itself. We owe to ourselves, to our neighbors around the world, to future generations and to the rest of life on earth to become better-informed citizens, to reconnect with our local communities and the land where we live while at the same time broadening our awareness and understanding of this entire marvelous planet that God has entrusted to our keeping.

Conclusion

We cannot live on the earth without affecting our environment. For those of us tempted to get nostalgic, it is worth remembering that the rural countryside to which many of us hanker to return was nothing like the prehuman landscape. Even the ecology of the New World had been shaped for centuries by Native Americans before it was "discovered" by European settlers. The dreams of some radical environmentalists of a "world without us" may serve as an interesting thought experiment, but they are obviously of little help for living in the here and now—especially for anyone who values human life and culture and who takes seriously the biblical perspective on God's purposes for humankind. The important issue, of course, is *how* we live on the earth and affect the world around us. Will we or will we not choose to live in ways that sustain and promote the flourishing of our fellow human beings and the rest of life on earth? If we do, what will that look like? How will we deal seriously and in Christlike love with the severity of the challenges facing us this century?

We focus in the later chapters of this book on what the New Testament in particular teaches about the future of creation and the difference this might make to how we think about such questions. But of course the Old Testament too has a lot to say about how we live on the earth, both in its pronouncements about the value to God (and therefore to us) of nonhuman creation and in the principles that emerge in its regulations governing the way God's people Israel were to live on the land. For example, the jubilee law, which returned land to the original owners every fifty years, would hypothetically ensure that everyone in an agrarian society had access to a means of support for themselves and their family—to land. Even if someone fell on hard times and sold themselves as an indentured laborer to another person, eventually they or their family would be able to get back the land they lived on and to pass it on as an inheritance. We cannot know whether or not the jubilee laws were ever put into widespread practice in Israel, but the principle is clear: land is a gift from God, and no one can claim sole permanent ownership over it. The related institution of a sabbatical year, whereby land was given regular rest from crops every seventh year, similarly models not just the practical good sense of allowing land a regular fallow year, but also the God-ordained wisdom of regular times of rest and times for celebration of God's provision.

In the minds of Old Testament prophets there was a clear link between environmental degradation, the fact that the land is often not as fruitful as God intends it to be, and the failings of human beings in their care for it. For a people whose existence was bound to the land, there was no naivety about the potential of the land itself to suffer the results of human evil and injustice. The land was indeed fruitful and intended by God to provide bounteously for all, yet to reject God's purposes could mean drought and ruin and the suffering of creation itself. Often this is expressed as the earth or the land "mourning" (e.g., Is 33:9; Jer 12:4; Hos 4:3), an idea that we will see Paul pick up in the New Testament to describe the entire creation in its longing to be set free from its "bondage to decay" and in its "groaning as in the pains of childbirth" because of human sinfulness (Rom 8:21-22). There is a direct causal link drawn in Scripture between the breakdown of the relationship between humans and their Creator God and the breakdown of our proper relationship with the earth.[51] A pronouncement of Hosea's resonates particularly strongly in our own context:

> There is no faithfulness, no love,
> > no acknowledgment of God in the land....
> Because of this the land dries up,
> > and all who live in it waste away;
> the beasts of the field, the birds in the sky
> > and the fish in the sea are swept away. (Hos 4:1-3)

Scripture sets out clearly the way in which human behavior has the potential either to enhance or to harm and even destroy the very environment in which we live. It warns us against the folly of presuming that the earth is limitless or that God would not allow us to suffer the consequences of our poor treatment of his creation. But the Bible also sets out clearly, as we will see in subsequent chapters, the sure and certain hope that we have in Christ for restoration and a setting of all things right in the new creation.

GLOBAL CLIMATE CHANGE

Of all the changes facing planet earth, we focus in the next chapter on the single issue of global climate change. This is partly because climate change causes, feeds into or exacerbates many of the problems discussed above. It is also a major ingredient in many so-called natural disasters. For example, rising sea levels will make flooding by tropical storms and hurricanes and the effects of tsunamis worse. Floods by far cause the biggest loss of life, even more than volcanic eruptions or earthquakes. Changes in global weather patterns and the increased amount of water vapor in the atmosphere in a warmer world are likely to lead to more extreme weather events, whether of rainfall, droughts or heat waves. Climatic change is likely to cause the extinction of organisms that cannot easily migrate, and it will change patterns of where crops can be cultivated. Increasing carbon dioxide levels may acidify the oceans, thus exacerbating climate change by killing many of the organisms that normally absorb carbon by building shells and skeletons. Global climate change is a major issue facing humankind that truly affects everyone on this planet, wherever they live. Yet it is also a problem that has been caused primarily by the way we humans choose to live. For that reason alone it merits detailed discussion.

3

Global Climate Change

Warming of the climate system is unequivocal.

IPCC FOURTH ASSESSMENT

Truly I tell you, whatever you did for one of the least of these brothers and sisters of mine, you did for me.

JESUS (MATTHEW 25:40)

◆

SO WHAT IS THE PROBLEM?

The average global temperature has increased by about 0.7°C relative to the twentieth-century average. That hardly sounds like anything to worry about, does it? The increase sounds a bit greater in the United States, since Americans still habitually use the Fahrenheit scale—but even 1°F does not seem like a lot. It is understandable, then, that many people switch off at this point. Why get excited about such a modest temperature increase, especially when we know that the earth's climate has always varied between colder and warmer periods? For some, the prospect of a slightly warmer environment actually sounds quite enticing.

There are, however, at least three reasons why scientists who study the earth's climate have in fact become extremely concerned—so concerned that some of them have even dropped the usual reticence of scientists about getting involved in policy debates and have issued strong public warnings about what may lie ahead. Their concerns stem from (1) their increasing certainty that the temperature rise we have witnessed thus far is but the beginning of a period of increasing temperatures that will take us into a climate regime humankind has never before experienced; (2) their knowledge that increasing average global temperature also increases the frequency of a wide variety of weather extremes that are damaging or even catastrophic for the communities affected, all the more so given the current vulnerability of so many of the world's ecosystems and of many places with dense human populations; and (3) their observation that, although the planet has experienced a wide range of temperatures throughout its geologic history, the *rate* of temperature increase we are experiencing now is unprecedented and makes adaptation all the more challenging.

We discuss below the physics of how greenhouse gases can affect the climate, but for now it is worth noting that the current consensus among climate scientists is that even if we began immediately to reduce our production of greenhouse gases as fast as is conceivably possible, we have already committed the world to a temperature increase of about double what has already occurred. That is because it takes between several tens and hundreds of years for carbon dioxide to be removed from the atmosphere into the soil or the oceans. The sea level will also continue to rise for many centuries to come, simply because it takes that long to circulate into the deep part of the oceans the heat that causes expansion of the water.

Even if the temperature increase were to reach 2°C or more, is this anything serious to worry about? Most people who have investigated this think it certainly is. Although the much-heralded Copenhagen summit of almost all the nations on earth in December 2009 failed to achieve any consensus about what to do about global warming, it did recognize that climate change is one of the greatest challenges facing the world today. Perhaps the most positive thing to come out of that conference, the Copenhagen Accord, proclaimed that actions should be taken to keep any global temperature increase to below 2°C.[1]

In fact, one of the impediments to progress at the Copenhagen summit was that a significant group of nations, including many low-lying Pacific islands, were pressing for an agreement to allow a maximum temperature increase of only 1.5°C. That was because they could see clearly the dangers ahead, not least the submergence of their own nations under rising sea levels. In response to those concerns the Copenhagen Accord agreed to review the feasibility of a 1.5°C goal by 2015. Such goals are for the time being almost meaningless, however, since the world's nations were unable at Copenhagen to subscribe to *any* legally binding commitments for curbing greenhouse gas emissions. A careful study of the voluntary national emissions targets that have subsequently been offered suggests that, even if these are adhered to, they will lead to a temperature increase of much more than 2°C, and indeed might well lock the world into an increase of over 3°C.[2]

Such temperature increases are likely to lead to radical changes in agricultural productivity around the world, and eventually to a widespread reduction in the amount of food produced. This could well lead to increasing deaths as a result of malnutrition and impaired resistance to disease, if not starvation. As heat waves become more common, and hotter, many will die, particularly the very old and the very young. There have already been harbingers of that. During five days in July 1995 in Chicago, an estimated 700 people died as a result of a heat wave.[3] Even worse, in the first two weeks of August 2003, an additional 22,000 to 35,000 people died in northern Europe during a period of extremely high temperatures.[4] These events both occurred in high-income countries, though in common with other disasters it was mainly the poor, the elderly and the disadvantaged who suffered most.[5] In low-income countries the outcome is likely to be worse. The drought conditions associated with the European heat wave of 2003 also led to crop losses of around £8,000 million, which caused increases in staple food prices. Forest fires in Portugal that year were responsible for a further £1,000 million in damage. Though we rightly consider such financial losses to be less important than human losses, they nevertheless can have profound impacts on human welfare, especially in low-income countries.

The volume of greenhouse gases in the atmosphere now is higher than at any time since humans first walked on this earth. Indeed, the levels of the prominent greenhouse gases carbon dioxide and methane are both far higher than at any time in the last 420,000 years, a period for which we know the atmospheric record well from bubbles of air preserved in annual snowfall layers recovered from ice cores in Greenland and Antarctica.[6] The atmospheric record from the ice cores shows that the air temperature is closely correlated with the concentration of greenhouse gases. There is no physical reason why the earth should not become much hotter than it is now. Indeed, the geological record shows that it has done so from time to time in the distant, pre-human past. So there is no reason to be complacent about the circumstances we are in now. There is no reason to believe that somehow the problem will go away and that temperatures will return quickly to those they were before we started pumping huge quantities of greenhouse gases into the atmosphere.

Perhaps more worrying than the average global temperature increase itself is the rate at which the temperature is increasing. This is indeed unprecedented, even in the extremely long geological record. The world's plant and animal species simply may not be able to cope with the rate of change; and even those species that may otherwise be able to migrate or shift toward the poles and higher elevations are in many cases unlikely to be able to do so in time, especially given the fragmentation of the world's natural habitats. Human societies too are of course very vulnerable to a rapidly changing climate.

Human cultures and societies are built on real estate. We need land from which to extract mineral and energy resources, land for water, land for growing food, land on which to build dwellings. Apart from dwelling space, these requirements do not change as societies become less rural; it is simply that the people and the resources on which they depend are increasingly separated. That is why almost all wars are fought over the possession of land. And it is why it is possible that if rapid climate change makes some areas uninhabitable, as is likely, due to such things as drought, flooding and rising sea levels, then wars may be a consequence. What is certain is that there will be increasing numbers of refugees if rapid climate change continues. People have to live somewhere. High-income countries such as in North America and Europe may in fact initially feel the effects of climate change most strongly in pressures from refugees wanting to immigrate. This will present a real moral challenge: since it is the high-income countries historically who have been largely responsible for causing rapid climate change, how can they refuse to help those from low-income areas such as sub-Saharan Africa and southeast Asia who will suffer the consequences most?

Arctic Sea Ice

One of the most striking consequences of global warming can be seen in the Arctic: the sea ice extent and volume are decreasing dramatically (see figure 3.1).[7] Since 1979 satellite images have made it possible to measure the extent of sea ice accurately. The ice goes through a yearly cycle of freezing in the winter and melting in the summer, with the minimum extent being reached in mid-September every year. As this chapter was being written, the smallest sea ice extent ever had just been recorded in September 2012, with the six lowest September ice extents all having occurred in 2012 and the previous five years. This represents the loss of 1.1 million square miles of ice, which is half of the 1979–2000 average ice cover of the Arctic Ocean. To put this into

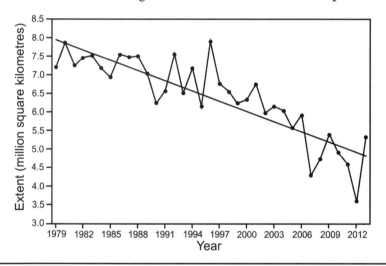

Figure 3.1. Average monthly Arctic sea ice extent for September 1979–2013.

perspective, the amount of ice lost is 140 times the area of Wales, or four times the area of Texas. The average rate of decline for September ice cover has reached 13 percent per decade relative to the 1979–2000 average.

Although the sea ice cover increases in area during the winter months, the new ice is much thinner than that which did not melt. So each year it becomes easier to melt back the one-year-old ice and to encroach farther on the older ice. Ironically, the decrease in Arctic ice cover is leading to an increase in oil exploration in the Arctic, which, if it results in yet more hydrocarbon extraction from this hitherto inaccessible region, will itself exacerbate global climate change.

WHAT CAUSES GREENHOUSE WARMING?

In order to understand the mechanisms behind climate change, it is helpful to picture the earth's atmosphere as a blanket that keeps the surface of the earth much warmer than it would otherwise be.[8] The effect was likened by the French scientist Jean-Baptiste Fourier in 1827 to the way in which a greenhouse keeps the inside warmer than the outside. A greenhouse works by allowing the sun's visible radiation—light—to pass through the glass unimpeded. The sunlight warms up the plants and the soil, which of course do not emit visible light but instead emit thermal energy that you cannot see with the eye. It is that thermal energy being emitted that you feel when you sit on a warm rock, or on dry earth, on a summer's evening after the sun has gone down. The glass of a greenhouse traps and absorbs some of the thermal radiation emitted by the warm plants and the soil. Some of that heat trapped in the glass is then radiated back into the greenhouse, thus helping to keep the inside warm (see figure 3.2). This basic idea of a greenhouse effect on earth has been known for nearly two centuries.

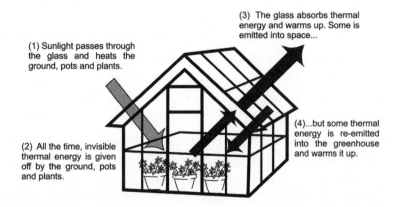

Figure 3.2. The greenhouse effect

The mechanisms behind this effect were elucidated more precisely by the British scientist John Tyndall, who in about 1860 was able to measure the amount of infrared energy absorbed by carbon dioxide and water vapor, the two most important greenhouse gases in the atmosphere. They play a role equivalent to the glass in a greenhouse. These gases have been present in the atmosphere, in varying concentrations, since life began. If it were not for these natural greenhouse gases, the earth would be more than 20°C (36°F) colder than it is. So instead of having an average surface temperature of about 15°C (59°F), the earth would be encased in ice. Life as we know it would be untenable. We can, therefore, be grateful for the greenhouse effect! The question for us today, however, is what happens when, through the burning of fossil fuels, we increase the concentrations of greenhouse gases.

As early as 1896, the Swedish scientist Svante Arrhenius made some theoretical calculations that suggested that doubling the atmospheric concentration of carbon dioxide would increase the average global temperature by 5 to 6°C. That value is not so far from our present estimates of a 2 to 5°C temperature increase made using complex computer simulations of the effect of doubling the amount of carbon dioxide in the atmosphere.[9] The basic mechanisms behind the greenhouse effect, then, have been understood since the nineteenth century and were worked out, appropriately enough, by international scientists.

Throughout its long history, the earth has maintained a delicate balance between the amount of carbon dioxide in the atmosphere and the temperature at the surface. The exact interplay between the two is complex and not fully understood, though it seems likely that life itself plays a crucial part in the balancing act. For the last 3.5 billion years the average surface temperature of the earth has remained between the freezing point (0°C) of water and its boiling point (100°C). This has happened despite the amount of heat coming from the sun having increased over the same period by about one third. If the surface temperature had dropped below freezing point for long periods, life could not have flourished. If it had ever increased above boiling point, all the water in the oceans and lakes would have boiled off and been lost to space, and the earth would have become a sterile place.

This complex self-regulating system should not be taken for granted. It operates on long timescales, much longer than a human life. For example, it takes many decades to remove excess carbon dioxide from the atmosphere by absorbing it in plants, soil and oceans. If there is an abrupt increase in atmospheric carbon dioxide, about half of it will be removed in thirty years, another third within a few centuries, but the remainder not for many thousands of years. The problem we face now is that humans have become agents of extremely rapid change on the earth. We are pumping greenhouse gases into the atmosphere at an unprecedented and truly astonishing rate.

By burning huge quantities of hydrocarbons—coal, oil and gas—we are releasing vast amounts of carbon dioxide. These hydrocarbons have been accumulated and stored underground over almost unimaginably long periods. For example, the oil and gas we are extracting from beneath the North Sea have been underground for over 150 million years, and yet we have managed to remove most of the easily tapped North Sea oil and burn it in just forty years. At its peak in 1999 some 250 million gallons (1,140 million liters) of oil were extracted from beneath the North Sea every day; and this is still less than one-tenth of the daily consumption of oil worldwide, and only 4 percent of the total volume of hydrocarbons burned every day.

The rate at which we are burning hydrocarbons is staggering. Ultimately all the hydrocarbons, be they coal, oil or gas, were produced by photosynthesis and growth of ancient vegetation, followed by its burial underground over geological timescales. So in a real sense we are using stored energy from the sun. It has been estimated that every gallon of gasoline we use required the growth of over one hundred tons of ancient plants. The current annual usage of hydrocarbons worldwide is equivalent to over five hundred years' worth of global plant productivity.[10]

The consequence of burning all these hydrocarbons and releasing their trapped carbon dioxide is, unsurprisingly, a dramatic increase in the concentration of carbon dioxide in the atmosphere. This has been monitored over the past fifty-five years at an observatory high on Mauna Loa, Hawaii. The increase in carbon dioxide is unambiguous (see figure 3.3).[11] The observatory is located in the middle of huge areas of barren lava fields, so plant respiration does not complicate the daily measurements. Because the observatory is at an elevation of 11,150 feet (3,400 meters) in the middle of the Pacific Ocean, far from the main sources of most of the carbon dioxide produced by human activity, this record represents a well-mixed average of how we are polluting the atmosphere. Since preindustrial times the activity of humans has increased the volume of carbon dioxide in the atmosphere by 40 percent. And the rate at which we are pumping out carbon dioxide is still increasing year by year.

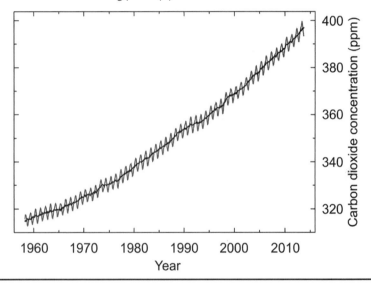

Figure 3.3. Increase in carbon dioxide concentration in the atmosphere (in ppm–parts per million) over the past fifty-five years recorded in Hawaii. The saw-tooth oscillations represent annual cycles, and the solid line the seasonally corrected average, which is rising inexorably.

Although carbon dioxide is the greenhouse gas that gets the most attention, there are many other gases that are also significant contributors to global climate change. Methane is the most important of these, producing nearly a quarter of the excess global warming, even though its concentration in the atmosphere is two hundred times lower than that of carbon dioxide. This is because methane is a much more potent greenhouse gas than carbon dioxide.

Methane comes from both natural sources such as swamps (hence its popular name of "marsh gas") and from the activities of humans. Growing rice, raising cattle, burning natural gas and mining coal have in the last century all added far more to the atmospheric concentration of methane than natural sources. Indeed, the methane produced by the 1.7 billion domesticated cows in the world accounts for nearly one-fifth of all greenhouse warming. In New Zealand, for example, 10 million cows and 40 million sheep produce nearly half of all the nation's greenhouse gases. These statistics remind us of the benefits of a lifestyle that involves eating less red meat than is common in the high-income countries, which, as we observed in the previous chapter, has other benefits for the earth and our health as well.

Several other gases, including chlorofluorocarbons (CFCs), are extremely powerful greenhouse gases. Molecule for molecule, CFCs are five to ten thousand times stronger greenhouse gases than carbon dioxide. Fortunately, their concentrations in the atmosphere are very low, though they still account for 12 percent of all the warming produced by greenhouse gases. CFCs also destroy the ozone layer, and the dangers of that led to the Montreal Protocol in 1987, when representatives of almost all the nations on earth agreed to ban production of CFCs by 2006. However, CFCs remain in the atmosphere for a long time—one to two hundred years—so it will take until 2050 before there is a significant reduction in their greenhouse effect. The Montreal Protocol has been ratified by 196 states, which led Kofi Annan, then secretary of the United Nations, to say that it was "perhaps the single most successful international agreement to date."[12] What we desperately need now is a similar cooperative and unambiguous international agreement on other greenhouse gases.

Are Humans Responsible for Climate Change?

The increase of the global average temperature over the past century is "unequivocal," in the words of the IPCC quoted at the head of this chapter.[13] The IPCC is a body that is very careful in its use of terms for statistical probability (and some have in fact accused it of being overly conservative), but it does not get any less ambiguous than "unequivocal." Each of the decades since the 1970s has been hotter than the previous one, including the first decade of the twenty-first century (see figure 3.4).[14] Temperature records are continually being broken. To take just one example, while Britain shivered under an unusually cold March in 2010, it was nevertheless the hottest March month ever recorded globally. This shows one of the problems with the public perception of climate change: it is hard to remember that the local weather we experience on a day-to-day or week-to-week basis is not a reliable guide to the global average trend.

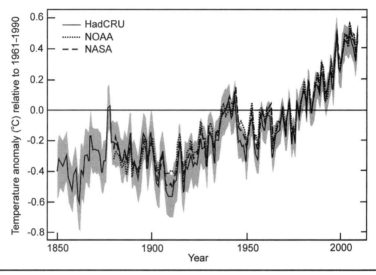

Figure 3.4. A clear correlation can be seen between three global average temperature records, which were created independently. Gray shading shows the range of uncertainty. They all show a marked warming trend, with each decade since the 1970s being warmer than the previous one.

If the actual climate change signal is so evident, the next important question is whether it is caused by human activity. Practically all scientists who have investigated this believe that it almost certainly is. However, it is fair to say that there is an outside possibility that the temperature changes are simply fluctuations in the climate caused by the normal processes of the earth system. But every year that goes by makes the influence of humans on global warming more apparent and renders other possible explanations less and less likely. We cannot afford to wait and wait before we take action. Even if by some chance the temperature rise were due to other causes, the changing climate would still be just as disastrous for many people and ecosystems. We have a good understanding of the physics of how greenhouse gases like carbon dioxide cause warming. So whatever the cause of the warming we are currently observing (though, again, nearly all climate scientists are now convinced it is due largely to human causes), reducing the volume of greenhouse gases in the atmosphere will serve to slow and reduce the warming.

If we look back through 420,000 years and four ice ages, we see a pattern of correlated changes in the temperature and in the greenhouse gases carbon dioxide and methane, all of which occurred before human beings impacted the climate (see figure 3.5).[15] On average, the temperature, carbon dioxide and methane all increase and decrease together. But we know that the ice ages are produced by cyclic changes in the orientation and orbit of the earth around the sun. Sometimes the earth gets colder as it receives less solar radiation, producing an ice age approximately once every 100,000 years. As the earth warms slightly and comes out of an ice age due to receiving more solar radiation, this rather weak warming is amplified by an increase in greenhouse gases (mainly carbon dioxide, methane and water vapor). As the large ice sheets in the northern hemisphere melt, this reduces the albedo (reflecting power) of the earth, thus reflecting less solar radiation back to space and causing yet further warming.

If we now look in figure 3.5 at the immense and abrupt increase of the greenhouse gases methane and carbon dioxide produced by human activity over the past century or so, it does not take much imagination or even complex computer programs to realize that the temperature is also going to increase. There is no doubt from the physics of greenhouse warming that if the volume of greenhouse gases in the atmosphere is increased, then the earth's surface temperature will rise. The record of the ice ages suggests that not only are the temperature and the greenhouse gas concentrations linked, but that increasing temperatures themselves lead to increased greenhouse gases in a cycle of feedback and amplification.

Other influences on climate change include changes in solar activity, which vary on a scale of years to a decade. El Niño and La Niña/Southern Oscillation events can influence global temperatures by creating bodies of surface water off the Pacific coast of South America with hotter or cooler temperatures, respectively. They last from about one year to a few years and occur typically about every five years. Volcanic eruptions can also in-

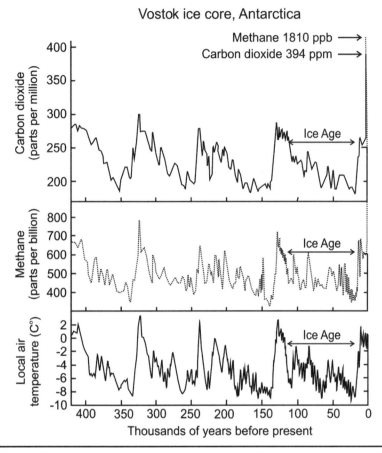

Figure 3.5. Variations in temperature, carbon dioxide and methane concentrations in the atmosphere over the past 420,000 years deduced from the Vostok ice core in the Antarctic. This period includes four prolonged ice ages, the last of which is labelled.

fluence global temperatures by injecting large quantities of dust and sulfur dioxide into the upper atmosphere, which reflect some of the sunlight and therefore cause cooling. However, their effect is temporary, usually lasting only a year or two, since the dust is rapidly removed by rain and snow.

What today's climate scientists have discovered is that when all of these known effects are put into computer climate models, they cannot reproduce the observed temperature changes of the last century. But when the greenhouse gases produced by humans are added, the agreement between the models and the observed temperature record is much closer (see figure 3.6).[16] This is a potent indication that humans are responsible for the rapid temperature increase since preindustrial times.

In summary, there is a strong consensus among climate scientists that climate change is real, is caused by humans and is potentially dangerous. Yet against that background, we all know that there is widespread doubt or even denial of the fact of climate change in the population at large. In the

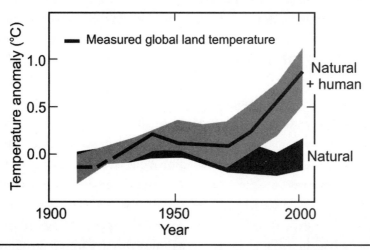

Figure 3.6. Measured global temperature variation over land in the twentieth century, compared to the results of multiple computer simulations of climate change using both natural forcings and man-made greenhouse gas emissions. Dark-shaded band: results of multiple computer simulations of climate change that include only natural factors. Light-shaded band: simulations including both natural factors and man-made greenhouse gas emissions.

United States, a survey in January 2010 by the Pew Research Center found that dealing with global warming ranks at the bottom of the public's list of priorities: just 28 percent considered it a top priority, the lowest measure for any of the twenty-one issues, including the economy, terrorism, healthcare and moral decline, tested in the survey.[17] Less than half the people polled believed that global warming was even happening. Things are not dissimilar in the United Kingdom, where a survey in February 2010 found that one-quarter of adults did not think that global warming was happening, and only one-quarter believed that climate change was real and "now established as largely man-made."[18] Perhaps more discouraging still was the finding that the levels of doubt had increased over the previous six months, a period that included the widely reported Copenhagen summit in December 2009 that sought to find global consensus on ways of dealing with climate change. In a later section we will discuss why there is such widespread apathy, skepticism or even denial of the reality and impacts of climate change.

What Is the Likely Future Climate?

We would all like to have certainty about the future. In reality much in human affairs is contingent and uncertain, both at the level of individuals and at national and international levels. What, then, can we say about the future climate? The answer is that it is surprisingly difficult to model in detail exactly what will happen at any particular place or time in the future. That is because on the ground we experience the weather on a day-to-day basis, and the weather often depends on small-scale local features. It varies a great deal. Computers are simply not big enough or fast enough to model the entire earth at the very fine resolution required to forecast the weather reliably a long way into the future. Added to that are the complexity of so-called feedbacks, about which we know little. The earth has never been this hot in recent experience. So we do not know, for example, whether the increased acidity of the oceans caused by a rise in carbon dioxide will prevent some of the marine organisms that normally absorb carbon into their shells from doing so in the future. That could cause a positive feedback, which means that carbon dioxide in the atmosphere would rise even faster because less of it would be absorbed into the ocean.

We can, however, make some educated guesses at what the climate will do. A warmer world also means a wetter world. More water vapor in the atmosphere is likely to lead to more extreme weather patterns. Storms are likely to be stronger, rainfall heavier in some places, droughts more prolonged in others, heat waves hotter and more common in others. Some places indeed will suffer the double whammy of drought one year followed by devastating floods the next, both of which will make growing crops sustainably extremely challenging if not impossible. Because of this local variability in climatic response, some people find it more helpful to talk about "global climate change" than about "global warming," though of course behind all these changes is in fact a globe that is overall getting warmer and warmer.

Global Climate Change 69

The best estimates of future temperatures up to the end of the present century are shown in figure 3.7.[19] There is a large range in predictions, from 1.5 to 6°C (depending in part on our own collective choices), although the best estimates are in the range 2 to 5°C. Those are quite high enough increases to be scary. Some of the potential consequences of increased temperatures are explored by Mark Lynas in his book *Six Degrees*, which won the Royal Society award for the best popular science book in 2008.[20] Many of those who have investigated the impacts of climate change believe that any increase above 2°C would be potentially catastrophic, but unfortunately we already look set to exceed that threshold.

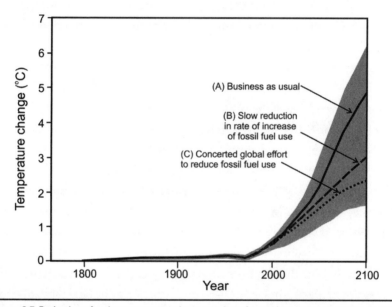

Figure 3.7. Projections for three representative scenarios of greenhouse gas increases as projected from multi-model estimates with climate models. Shading shows the range of uncertainty.

We are in uncharted territory as the temperature rises get higher and higher. But we can make educated guesses about the impact in a variety of areas such as food production, the health of ecosystems and of human beings, water availability and coastal vulnerability. Many of the consequences will lead to increased numbers of refugees. Probably hundreds of millions of people will have to move out from areas made inhospitable by extremes of climate, failure of agriculture, flooding or loss of water as temperature rises become extreme. For temperature increases of 2 to 3°C, agricultural yields in tropical and subtropical areas are likely to decrease, but—with adaptation—those in mid and high latitudes could remain stable or even increase. However, as average temperatures rise through 3 to 6°C, even the temperate regions become poorer for agriculture as summer temperatures soar and pressures increase on water supply. The total potential food production globally will then begin to decline.

Ecosystems, many of which we have seen are already in trouble due to other causes, will be increasingly pressured as temperatures rise. Even for a 2 to 3°C rise, one-third of all species will be under threat of extinction, corals will begin to die and there will be increasing risks of major wildfires. With a 5°C or more increase, there will be devastating extinctions around the globe, including widespread coral mortality. Much of the terrestrial biosphere will actually become a net contributor of carbon dioxide to the atmosphere, rather than absorbing it as at present, thus further exacerbating the warming of the planet.

Changes in water supply will become pressing especially at the higher end of temperature predictions. As glaciers melt they will cease to provide a stable year-round supply of water, affecting billions of people worldwide. For example, one billion people currently rely on water supply from the Himalayan glaciers. More extreme floods and storms coupled with rising sea level will cause the loss of about 30 percent of coastal wetlands for a 5°C increase, and many low-lying deltaic areas and islands will become uninhabitable. Health effects are difficult to quantify, but in addition to the pressures of high summer temperatures, which may be fatal for vulnerable people such as the very old and the very young, there is likely to be a general increase in diseases related to greatly increased temperatures. Diarrhea, malnutrition and respiratory complaints will disproportionately affect the poorer and marginalized in our societies, especially in low-income countries. Inevitably these will lead to increased deaths that otherwise would not have occurred.

It is difficult to predict what adaptive measures might be possible and feasible to counter these threats, but there is no doubt that it is the poor who will suffer most. Given the grim scenarios for temperature rises at the upper end of the predictions, there is certainly a strong case to be made that avoiding the problem of large global temperature rises in the first place is by far the better option than letting them happen and then hoping we (or rather our successors) will be able to cope. Yet the world stubbornly persists in following the upper limit of all the greenhouse gas emission scenarios under a business as usual pattern.

The actual temperatures that occur will depend on many imponderables. One is the uncertain feedbacks we have already mentioned. Another is the size of the global population. The global population quadrupled in the twentieth century alone. All else being equal, doubling the population will roughly double the production of greenhouse gases. Of course, not all societies are equally polluting. A child born today in North America will likely produce fifty times the lifetime total of greenhouse gases of a child born in India or sub-Saharan Africa. Nor can we be certain about the degree to which the world will be able to reduce its production of greenhouse gases. The predictions in figure 3.7 attempt to model various realistic scenarios. But one thing is very apparent. Thus far we have been running along the worst-case scenarios for production of carbon dioxide. Despite the growing awareness of the dangers of global climate change, and abundant political rhetoric, so far nothing much has changed. We are still polluting the atmosphere at the upper limit assumed in all the models, with business as usual.

Tipping Points

There are two important aspects of global climate change that are often overlooked. The first is that the climate responds only sluggishly to changes. It is not unlike trying to turn around a supertanker: it can take three miles to make a supertanker come to an emergency stop with the engines in full reverse. What we have already done in polluting the atmosphere means that we have committed the world to continuing changes for decades and centuries to come. Action to decrease our greenhouse gas emissions now will not begin to have an observable effect until the second half of this century. Temperatures are certain to continue rising until then, even if we were able to hit the emergency stop button right now (which of course we cannot). So our children and grandchildren will undoubtedly have to deal with the consequences of our actions.

The second aspect is that we may cross tipping points from which there is no return. An example would be complete melting of the summer Arctic sea ice, which, as we discussed earlier, has already lost half its area since satellite observations began in 1979. Since ice reflects sunlight, whereas water is much darker and absorbs heat, this produces a positive feedback. As the ice melts, the water heats up more than it otherwise would have done, and it can cause a runaway effect. The loss of the sea ice acts as an amplifier which makes the temperature increase even higher. It is estimated that if the planet warms up by an average of 2°C, the Arctic region would warm up by about 5°C. A tipping point such as this happens when a small perturbation is capable of driving rapid change. It is possible that we have already crossed the critical threshold for losing the Arctic sea ice.

Another high-risk possibility in the same region is that the Greenland ice cap could undergo runaway melting. The warm water from partial melting is apparently already lubricating the glaciers that flow into the sea and allowing them to flow and then break away as icebergs much faster than in the past. The consequence of melting all the Greenland ice would be a two to seven meter global rise in sea level, which is extremely significant. The timescale for this is slower than the melting of the Arctic sea ice, taking perhaps three hundred years before all the Greenland ice would be gone. As with all tipping points it is hard to know exactly when they will kick in, but for the Greenland ice cap it is estimated that the tipping point, after which all the ice would inevitably melt, will occur at an approximately 3°C rise in temperature. So there should still be time to avert it, provided we take concerted action to do so now.

A range of other possible tipping points have been discussed by a group of senior climate researchers,[21] but given the complexity of the interactions among the atmosphere, the climate and the surface of the earth, there may well be others that have not yet been recognized. The possibility of tipping points with major uncontrollable consequences means that it is important to consider mitigation strategies to minimize as far as possible the production of greenhouse gases in the future, as well as helping those most affected by climate change to adapt to changes that are already unavoidable.

ARE THERE TECHNOLOGICAL SOLUTIONS TO CLIMATE CHANGE?
Repeatedly in the past, humankind has gotten itself out of trouble through technological solutions. This is often on display in times of war, when a nation mobilizes all the ingenuity and resources at its disposal to deal with a threat that is considered worth the cost of facing head on. More positively, we have seen that the exponentially rising human population of the past two centuries has not caused a Malthusian disaster of worldwide famine and starvation. Instead, global food supplies have increased through the mechanization of agriculture and the Green Revolution. This may well pose serious challenges to the long-term sustainability of agriculture, but in the short term it has provided great gains in food production. Current research in pest- and disease-resistant genetically modified crops similarly remind us of the potentially major benefits (and challenges) that attend technological innovation.

It is in any case likely that new technology will have to be part of any "solution" to climate change. Energy supply is of course one of the main areas in which technology can play a role. If we are to decrease our production of greenhouse gases while maintaining or increasing our energy production, for example, we will need to invest heavily in methods of using renewable sources (wind, tide, sun, geothermal and hydro) to generate electricity; we will have to move toward an element of local energy production at the individual or community level; and, more controversially, we may, some suggest, have to use nuclear energy as part of the mix.[22] And if we are to continue to burn coal and natural gas for energy, which seems likely given the large supplies still available, then we will need to perfect ways of capturing the carbon dioxide produced and storing it underground without letting it get into the atmosphere.

All of the above are possibilities, though they will cost substantial amounts of money to implement and certainly are more expensive in the short term than simply carrying on as we are doing—burning coal or gas and venting the waste gases to the atmosphere, and using cars for personal transport at the level we do now. The other side of the coin, of course, is to find ways that enable us to use less energy. Such things as improving the insulation of houses, changing to low-power light bulbs and using our cars less are all relatively obvious things to do, but nevertheless may require that we rethink our lifestyles and be willing to make modest—and sometimes more radical—changes to the ways we arrange our lives.

What might we do if all the above are insufficient? Some people are beginning to talk about massive planetary-scale intervention, such as the possibility of spraying particles into the stratosphere to reflect some of the sunlight and so reduce surface temperatures, or seeding the ocean in an attempt to increase biological productivity and aid the removal of carbon into marine organisms.[23] Such last-ditch attempts to avert temperature rise are frightening to contemplate. They are likely to have dramatic and unpredictable side effects and may well produce cures that are worse than the disease. Nevertheless, it is probably important to start research into them now and—if appropriate—to attempt small-scale experiments simply so that we know a bit more about such options. The problem with all the interventions proposed thus far is that they would almost certainly have profoundly negative effects for some portions of the planet, even if on balance they had the positive effect of slowing the earth's warming. Who would control the decisions to launch such efforts? What provision would be made for the people and lands affected? At the moment, the uncertainty and the likely human and environmental costs of any such interventions are far, far higher than the costs of reducing our emissions.

Climate Change Skepticism

One of the most intriguing aspects of climate change for sociologists is that despite the depth of agreement among scientists that global warming is a reality and is happening right now, there is a large measure of skepticism and even outright denial by a large proportion of the public. Why should this be?

One reason is that unless we are hit by some catastrophic event such as prolonged drought or devastating flood, the year-by-year changes experienced by an individual are much smaller than the normal variability in the weather. Our individual experience is of the vagaries of the weekly weather rather than the long-term effects of climate change. Although global changes are currently happening at a historically unprecedented rapid rate, nevertheless they are on timescales of decades rather than years. So unless we are alert to the scientific data, we are not likely to consider such changes as important. This is especially the case if taking such changes seriously might seem to require fundamental lifestyle changes and expenditure of money. So we simply accommodate ourselves to a world in which the conditions are a bit different from how they were before—provided, that is, that we live in an area where the changes are not yet dramatic and that we have an income sufficient to enable us to adapt easily.

The other challenge, however, is that even for those who are open to considering the evidence, for most of us our information on such matters comes, of necessity, from the media. One of the ways the popular media keeps our attention is by highlighting the unusual, the extreme and the controversial. A 0.1°C rise in average global temperature over the past decade is not high-impact news. Possible fraud is. So when hackers in November 2009 illegally obtained copies of private emails between climate scientists from the University of East Anglia and elsewhere and found a few with slightly racy language—the sort of thing frequently heard over a pint in the pub—they cried "conspiracy," "fraud," "cover-up." The media publicized it massively, and the public's belief in the trustworthiness of climate scientists plummeted overnight.

It is certainly the case that scientists, like everyone else, are fallible human beings. They can and do make mistakes and sometimes frustrations or rivalry can get in the way of dispassionate observations. But the whole nature of the scientific method is that it is self-correcting. The foundation of the scientific method is its repeatability. Scientists are by nature skeptics. If someone comes up with unusual results, people in other laboratories will conduct experiments to either replicate or disprove them. One of the surest ways to make your reputation as a scientist is to overturn received wisdom on some theory and replace it with a theory that explains the data better.

In the case of the emails hacked from the University of East Anglia, the handful of scientists concerned were cleared by successive enquiries of any major wrongdoing, apart from sometimes intemperate language in a few private emails, which apparently arose from frustration. No matter that the global temperature curves to which the scientists had contributed were indistinguishable from those constructed by other groups working independently and with different datasets (as figure 3.4 demonstrates). No matter that their published work was all rigorously peer-reviewed by fellow scientists. It became a news item and, presumably on the basis that "there is no smoke without fire," the public's confidence in what all climate scientists said decreased.

This is, of course, a crazy way to conduct public debate. It is exacerbated not only by the media's need to find striking or controversial news to keep the attention of their audience, but also by the way the media almost always tries to find two people of opposing positions to debate these issues. So even if 99 percent of scientists believe that global warming is real, the radio listener or television watcher will often hear one person saying that global climate change is happening while another says it is not. To the listener it sounds as if the positions carry equal weight and that there is much less agreement on the issue than is in fact the case.

Another partial reason for lack of clarity about climate change among the public is that scientists have often been rather poor at explaining their results in lay terms. They usually prefer to have their heads down analyzing the data than doing public relations and explaining their results to nonspecialists. There has also perhaps been a tendency to hubris—the idea that scientists are the experts and know all the answers. Both of these things must change, because the public needs to own the results and to be carried along with the scientists if they are being asked to sign up to hard and perhaps costly changes.

Perhaps the most important factor in the public's uncertainty about the reality of climate change—and a factor that has had a major influence on the media's reporting of climate change—is the deliberate and quite cynical efforts, mainly by certain large corporations, to downplay the possibility and results of climate change in order to maintain their own profitable industries. These activities, funded to the tune of many millions of dollars by giants such as ExxonMobil, have been well documented.[24] The strategies adopted by these bodies have been as follows.

- To raise doubt about the scientific evidence, even when it is well attested by all normal scientific methods and by unbiased refereeing.
- To impute as politically biased or unsound the reports of the influential IPCC, which represents the consensus view of thousands of scientists from all around the world.
- To attempt to portray its opposition to action as wisely waiting for certainty in the science rather than as the naked self-interest it actually represents. It is worth noting that identical delaying tactics were used by the tobacco industry to downplay acceptance of the scientifically clear link between smoking and cancer. That delay cost many thousands of lives, and it is likely that delay in dealing effectively with the causes of climate change will likewise cost thousands or, more likely, millions of lives.
- To emphasize the cost of mitigation and to downplay the cost of the damage likely to be done by climate change. Yet, as Nicholas Stern's 2007 report showed,[25] it makes good economic sense to produce fewer polluting greenhouse gases now, because in the long term it will save far more in the extra costs that would otherwise have been incurred. Investing in alternate energy sources is likely to be a wise move in its own right, making the consumer less vulnerable to increasing prices as oil reserves are depleted and decreasing their nation's dependence on oil from often unstable parts of the world.
- To point to inconsequential uncertainties or errors in reports that do not actually change the overall picture one jot. Blackening the reputation of the University of East Anglia climate scientists is included in this category, since whatever they may or may not have done or written, it makes absolutely no difference whatsoever to the observed trends of increasing global temperature. Yet the public may think that it does.
- To conflate uncertainty (which is a quite proper scientific measure of predictions) with the idea that this means scientists think that global change may not in fact be happening.

Though the number of climate change skeptics may be small, their influence is large, partly because they shout louder than those who accept the consensus scientific view. It is also partly because we all love to hear a message that means we do not really need to change anything we do. This brings us to our next question: Why is it so difficult to make the sort of changes that are required if we are to attempt to mitigate climate change?

Why Is It So Difficult to Make Changes?

A curious and disquieting feature of the global climate change debate is that despite its importance ultimately to every single citizen of this planet, to their children and their grandchildren, and its global reach and likely severe consequences for health, for agriculture, for international stability and for the future of life on earth, it is extremely hard actually to make the changes that are necessary to slow global warming. We have already mentioned one of the pertinent reasons: we will suffer the consequences of climate change some way into the future and it is hard to put effort into the changes required when we have our daily lives to live. Our governments struggle too, for not dissimilar reasons. Most democratic governments are elected for terms of only four or five years, so it is hard for them to make costly changes when the benefits may not be seen for years or, more likely, for decades into the future.

There are two ways we might reply to this. The first is the self-interest argument, which was put forward by Nicholas Stern in his 2007 report. From purely economic arguments alone, he reported that it is worth investing as little as 1 to 2 percent of Gross Domestic Product (GDP) each year into a low-carbon economy because it could save up to 5 to 20 percent of GDP in the future. There is some debate about the exact value that should be used for discounting the future, which changes the magnitude of the amount that would be saved in the future, but the general principle holds true.[26] Sooner or later we will in any case have to wean ourselves off oil, since most analysts agree that we have already burned about half of the world's reserves. There is even an economic driver here: a company that is early into low-carbon products stands to make large profits when everyone else has to buy them from necessity.

The second response is the one that we develop in the remainder of this book. If we are truly to love our neighbor as ourselves, as Jesus commanded us to do, then those of us in the high-income countries that historically have caused global climate change through our emissions of greenhouse gases have to take account of the effect of our actions on our neighbors and on all of life on earth. Those neighbors may be invisible to us, either because they live in faraway places such as southeast Asia or sub-Saharan Africa or because they are not yet born. Yet, given what we now know about the consequences of our actions, our responsibility toward them remains. We enjoy a high standard of living largely because of our burning of fossil fuels both today and over the past century or more. So we have a responsibility to help those affected by this to adapt to their changing circumstances and to do what we can—individually, communally, politically—to stop wreaking such damage, to try to prevent the worst-case scenarios climate scientists warn us about.

There is a palpable sense of gloom among many climate activists—a sadness and even despair that certainly does not help draw many others to their cause, but that is understandable given the severity of the challenges we face. Nonetheless, as the rest of this book explores, Christians have hope for a future that does not depend on us saving the planet. Christians envision, in fact, a whole new creation that is given by the grace of God. What does that mean for how we think about climate change and about all the challenges detailed in the previous chapter? What does it mean for how we act in the present? In the next chapter we consider how biblical visions of the future can be relevant to such questions, and we ask what any of this has to do with the gospel of Jesus Christ.

4

Why Hope?
The Gospel and the Future

The future is not a gift—it's an achievement.

ROBERT F. KENNEDY

*Do not be afraid, little flock, for your Father
has been pleased to give you the kingdom.*

JESUS (LUKE 12:32)

◆

QUESTIONING BIBLICAL HOPE

The previous two chapters have outlined some of the challenges that we and the whole earth are facing in the early decades of the twenty-first century. The next four chapters look at what the Bible, written thousands of years ago, has to say about the future of creation. This chapter addresses the question of how these two things can possibly be related. Even for Christians who take Scripture and the gospel seriously, what difference can something as seemingly obscure as New Testament eschatology (teaching about the "last things") make to how we think about and address today's environmental issues?

For starters, there are plenty of secular writers who think that our perspective on the future makes a lot of difference. Some consider that Christian notions about life after death have hopelessly muddled our thinking about life before death; they think it has made us incapable of responding properly to the challenges of the modern world. The bestselling author Ian McEwan, for example, uses some of Norman Cohn's popular arguments about the dangers of millennial belief to call on religious believers today to accept the fact that, even if God does exist, he does not and will not intervene in human history.[1] For McEwan, believing that God is actually involved in the world—and especially believing that he will act in final judgment—leads to paying insufficient attention to what is going on now. He thinks that it excuses people from trying to improve life, society and the environment. David Orr, an American political scientist and editor of the scientific journal *Conservation Biology*, similarly claims that the evangelical Christian's goal of "a redeemed world that fulfils the promise of creation" is irreconcilably opposed to the conservationist's goal of a sustainable world. Orr in fact blames this Christian ideology and conservative American evangelicals in particular for holding up conservation efforts in the United States and elsewhere.[2] The quote at the beginning of this chapter reveals the presupposition that lies behind the secular view: if people believe that the future is a gift, they have no reason to work to try to make it better.[3]

These arguments would not be surprising if they were made only by non-Christians. But Christian writers also sometimes admit as much, no doubt because they recognize that there is plenty of evidence that some versions of future expectation do in fact lead to present neglect. In an address calling on Christians to take up their responsibility to care for the earth, Rowan Williams, the former Archbishop of Canterbury, warns us against the folly of treating God's faithfulness like a "safety net that guarantees a happy ending in this world." He thinks that such a view of the future makes us prone to ignore the consequences of our sins against others and against the environment too.[4] Biologist R. J. (Sam) Berry, who has long taken the lead in efforts to encourage Christians to work toward environmental sustainability, similarly claims that pragmatic reasons force him to ignore "debates about the coming judgment and conflicting interpretations about the Millennium, Armageddon and the Apocalypse"; as Berry says, "they inflame passions and consume lives."[5] Theologian Kathryn

Tanner goes further and suggests a creative way of avoiding such debates entirely. Echoing attempts by Rudolph Bultmann in the last century to demythologize those parts of the Bible that do not fit with a modern worldview, Tanner proposes that we give up Christian eschatology and reframe it in purely spatial, nontemporal terms.[6] She suggests that rather than looking to the future for our hope, we ought to look only for the possibility of grace breaking into our lives in the present.

This returns us to a question that we raised in chapter 1: Why focus on what the Bible says about the future if it is so challenging, unpopular and even potentially damaging to an environmental ethos? We would seem to be on much firmer ground—and also to have plenty of biblical warrant—if we based our discussion solely on the ways in which Christian *love* demands that we care for the earth.

The theologian Stephen Williams has pointed out that no matter what Scripture says about the future, Jesus' command to love our neighbors as ourselves requires love to remain always at the center of our engagement with the world.[7] As the apostle Paul tells the Corinthians, of the three virtues faith, hope and love, "the greatest of these is love" (1 Cor 13:13). It has become increasingly clear that to love our neighbors today will require caring for the environment in which they live. So even if we concluded that the Bible had *nothing* to say about the future of life on earth, we would still have plenty of reason to engage seriously with the environmental challenges facing us. Nevertheless, there are at least two reasons why it remains important for us to attend to what the Bible says about the future of creation—and these two reasons are in addition to the urgent, practical need to provide a robust Christian response to today's apocalyptic environmental rhetoric.

Two Reasons Not to Give Up on Hope

The first reason to pay attention to Scripture's vision of the future is that the contours of biblical hope provide for us the context in which Christian love and charity are to be worked out. Some Christians have thought from time to time that the earth and all of its nonhuman life is intended only as a backdrop, stage or window-dressing to our personal relationship with

God—or even represents a material existence from which we hope ultimately to escape. If we think like that, then obviously the ways in which we relate to the nonhuman world around us are likely to be rather different than if we think that God also cares for the rest of creation and that he intends us to reflect this in how we express our love for him and for our neighbors.

For Christians, the inherent value of nonhuman creation, what is sometimes called its "intrinsic" value, derives in the first instance from God's pronouncements in Genesis that all that he has made is good. If God values what he has made (even before the appearance of humankind, according to Gen 1:4, 10, 12, 18, 21, 25), then we ought to value it too. How can we love God and not also come to love what he loves? The intrinsic value of a created thing—something that owes not just its material makeup but its very being and even possibility of existence to a Creator who "calls into being things that were not" (Rom 4:17; see also Heb 11:3)—is rooted in the value that is bestowed on it by that Creator. Its value is intrinsic, then, in the sense that it is not arbitrarily assigned to it nor can it be taken away by another mere creature. Its value cannot be changed by other beings who owe their own existence to the same Creator. This is why Christians sometimes talk about the "sacredness" of life. All of life is given by God and so derives its value in relation to him; it is not ours merely to do with whatever we please. A biblical theology of creation, as well as a biblically derived environmental ethos, is rooted in the affirmations in Genesis and many other passages in both the Old and New Testaments that it is God alone who establishes, upholds and sustains all of creation.

We cannot stop there, however, because that is not the end of the story. Many readers of Genesis, for example, understandably question whether God's pronouncements that creation is "good" or "very good" in the first chapter still apply after the third chapter, after human beings have rebelled against God and fallen under the curse that distorts their relationship with God, with each other and with the land itself. Have the effects of this fall so corrupted the earth and its life that God no longer cares for creation, or maybe has even decided to abandon his plans for it? We naturally want to read the rest of the Bible to find out what happens next, to learn whether even this present fallen creation, racked as it is by human sin, continues to be valued by the Creator. What place does this earth and its creatures and all of nonhuman creation have in God's ultimate purposes?

The goodness of creation as described in Genesis is sometimes explained as a fitness for purpose. This is a goodness that has to do in part with the way in which the created order fulfills, or begins to fulfill, God's intentions. It would be a mistake, then, to stop with Genesis chapters 1–2, to pretend that the fall of humankind never happened or that redemption is unnecessary. We cannot ignore what the rest of Scripture says about the ways in which creation itself is caught up in the drama of fall and redemption. Genesis and other biblical passages that describe the origins of creation provide the indispensable basis for any Christian discussion of how we relate to the rest of creation. But this is only the beginning of a story that draws us in and pushes us to go on searching—and even to yearn—for its resolution, its ending and its fulfillment.

This brings us to the second reason why we need to pay attention to what Scripture says about eschatology: the Christian gospel itself is driven by hope and is inseparable from its orientation toward the future. We will consider in a moment just what we mean by the *gospel*. But it needs to be stressed from the outset that if we are to look to Scripture to derive our faith and our practice, we are inevitably going to be confronted by what it says about those things that are yet to come. Despite the creative efforts of Kathryn Tanner and others to get rid of awkward Christian doctrines like the final judgment, resurrection and new creation, these doctrines are deeply embedded in the story that the Bible tells us about God and his relationship to his creation. Above all, these are not merely abstract doctrines but are the very elements that compose the story embodied in the life, death and resurrection of Christ himself.

Eschatology is an inescapable and even central part of biblical (and especially New Testament) thought because the outworking of God's faithfulness and righteousness is throughout the Bible bound up with the future vindication and full revelation of his divine justice and his love. Karl Barth accurately summarized the biblical perspective when he observed that, without hope, "there is no freedom, but only imprisonment; no grace, but only condemnation and corruption; no divine guidance, but only fate;

no God, but only a mirror of unredeemed humanity."⁸ According to the New Testament writers, this is the situation in which we all find ourselves; this is the plight of the whole world, the result of our sin cutting us off from God, the only source of our life and hope. Only in Christ is the situation overturned. It is in Christ that God's promises reach their fulfillment, in Christ that the future is secured and in Christ that the new creation begins to break into the old. As German theologian Christoph Schwöbel explains, "Jesus' story discloses the character of God's relationship to his creation as one by which God maintains his relationship to creation through the discontinuity of death," and "this story is a promise for all."⁹

The biblical narrative draws us into a drama in which the climax has been reached already in the death and resurrection of Jesus. It is here that God's purposes for us and his love for his creation are revealed definitively and absolutely. As those who have been given the joy and the privilege of entering into the story at this point, as those who trust in Jesus for our salvation and find our new identity in him, the way we live in the present is thus determined and shaped by what God has done already in Christ. But the final act is yet to come. And so our lives are also oriented toward all that God has promised is still to be. They are directed toward the future that has been secured for us through the incarnation, death and resurrection of the Son of God. This glorious future shines light back into our present, drawing us forward in expectant hope and calling us to become the men and women God intends us to be.

By attending carefully to what the Bible has to say about those things yet to come, we can learn something about the value of nonhuman creation within God's purposes. We can also learn where we fit in with those purposes. What sort of attitude ought we to adopt toward the future? What practices ought to accompany this attitude? If hope drives the Christian gospel and is inextricable from the story that the Bible tells us about God and his relationship to us and his creation, it will be impossible for us to live as gospel-centered people without looking toward the future promised us in Christ. This returns us, finally, to ask what it is that we mean by the Christian gospel. What does the gospel have to do with the earth and its future? This question is ultimately the one that drives this entire book, so before we go any further we ought to remind ourselves of just what we mean by *the gospel*.

THE GOSPEL, CREATION AND THE FUTURE

Christian ethics—including the ethics of how we care for the earth—ultimately derive from reflection on what it means for us today to live as citizens who are "worthy of the gospel of Christ" (Phil 1:27).[10] This is why we had better be sure that we have a firm grasp on what exactly "the gospel of Christ" means. It will do no good for us to spin theories about what a Christian environmental ethos ought to look like if we fail to root it in the story of what God has accomplished and revealed to us in and through Christ. In fact, the failure of many of us in the Christian church to take seriously our responsibilities within creation may be traced in part to our failure to grasp the significance of the gospel story in all its fullness. It is worth taking a moment, then, to reflect on this wider story in which the earth's future, and ours, finds its place.

If you were asked to summarize the Christian gospel, what would you say? Where would you begin? Where would you end? You might be familiar with one or two popular summaries of the gospel wherein the biblical story is distilled into a handful of easily remembered principles about how we as individuals relate to God. Perhaps you have seen, for example, the simple line drawings illustrating two ways to live or have heard evangelistic presentations of four spiritual laws. These sorts of presentations can be helpful for quickly and simply summarizing important truths about God and how we can be restored to relationship with him. They manage to convey in a very short space such things as the fact that we have become alienated from our Creator because of our rebellion against his rule; that we are unable by ourselves to bridge the gulf that now stands between us and him; that God has provided in Christ the means by which we can be restored to relationship with him when we place our trust in him; and that God frees us by his Spirit to serve him as our Lord and Savior. Simple, clear and powerful presentations of the significance of the gospel along these lines have helped many people—no doubt including some readers of this book—to come to know and to follow Christ.

For all the helpfulness of these summaries, however, they can be misleading if we fail to recognize their limitations. As abstract distillations of one aspect of how the biblical story applies to us as individuals, they necessarily miss the bigger picture. By focusing entirely on our personal relationship with God, they may lead us to forget the way in which that relationship is part of a wider web of relationships between the triune God, his people and all of his creation. By treating the gospel solely as a timeless truth for us today, they may cause us to miss the ways in which Jesus is presented throughout the New Testament as the fulfillment of the long-held hopes of God's people Israel. Above all, there is the danger that if we rely solely on isolated, abstract concepts to summarize our faith, we will miss the full significance of the incarnation and resurrection of Jesus. We risk making the gospel a story all about us rather than about God. None of this means that such clear, simple and concise approaches are without merit or cannot be used effectively in all sorts of situations. It would be a travesty if we stopped sharing the gospel with our friends and neighbors simply because we cannot always tell the whole story or tell it as well as we would like. But we do well to remember the limitations of such summaries and to remind ourselves that they are not the gospel itself, but only derivations from it.

Well, then, just what is the grander picture or wider story of the gospel? We will begin by looking at a handful of New Testament texts to see which elements the biblical writers themselves mention when they summarize the gospel. Paul provides us with the most relevant examples, so we will focus especially on how he interprets the gospel for his readers. But the proclamation of the good news goes back to Jesus himself and, as we will see, is anticipated already in the Old Testament.

According to Luke's account of Jesus' life, on a sabbath near the beginning of his ministry Jesus went into the local synagogue in his hometown of Nazareth. He unrolled the scroll of Isaiah and read from it to the assembled worshipers:

"The Spirit of the Lord is on me,
 because he has anointed me

> to proclaim good news to the poor.
> He has sent me to proclaim freedom for the prisoners
> and recovery of sight for the blind,
> to set the oppressed free,
> to proclaim the year of the Lord's favor."
>
> Then he rolled up the scroll, gave it back to the attendant and sat down. The eyes of everyone in the synagogue were fastened on him. He began by saying to them, "Today this scripture is fulfilled in your hearing." (Lk 4:18-21; citing Is 61:1-2)

What is the "good news" proclaimed by Isaiah that Jesus claims he is bringing to fulfillment? It is the arrival of the "year of the Lord's favor," the time appointed by God to bring healing, freedom and restoration. Through the power of the Holy Spirit, Jesus, the Anointed One, the Messiah, brings in the era foretold by Isaiah when God would act to save his people, when his kingdom would begin to be realized on earth. This rule is marked by physical and spiritual healing, by rescue from oppression, by restoration and (if we read right to the end of Isaiah) by new creation.

The rest of Luke's Gospel displays the signs of all this already in Jesus' ministry. We see it in such things as the "restoration" of a man's withered hand (Lk 6:10), the release of one "bound" when Jesus straightens a bent and crippled woman (Lk 13:12-16), and the many healings and inclusion of those who had otherwise been excluded—lepers, the demon possessed, a woman with constant bleeding. As Isaiah foresaw, God intends in Christ to address the physical, social, spiritual and cosmic consequences of humankind's brokenness and alienation from God. When John the Baptist's disciples come to Jesus wondering if Jesus is indeed "the one who is to come" (Lk 7:19), Jesus answers with an appeal to these marks of God's rescue mission that can be seen in his ministry: "Go back and report to John what you have seen and heard: The blind receive sight, the lame walk, those who have leprosy are cleansed, the deaf hear, the dead are raised, and the good news is proclaimed to the poor. Blessed is anyone who does not stumble on account of me" (Lk 7:22-23). The signs of God's kingdom are evident already in Jesus' words and deeds. The good news that is being proclaimed to the poor is that "God has come to help his people" (Lk 7:16).

Notice too, however, the implicit warning in Jesus' response, the hint that some might take offence at him. After all, John the Baptist is stuck in prison. How does this fit with the fact that Jesus is bringing in the kingdom of God, a time of rescue and restoration? For an answer, Luke's readers have to wait until the end of the book. There we learn that Jesus' mission involves more than merely healing and restoring those people whom he happens to meet over a few years as he travels through Galilee and into Jerusalem. Jesus' mission is to carry out God's radical plan to save his people—to save all peoples—from their sin and to make possible forgiveness, reconciliation, restoration, new life, new creation. This can finally be accomplished through nothing less than the cross, where Jesus, the righteous Son of God, is tortured and crucified. As Paul will spell out more fully, it is here that the world's evil is definitively dealt with, here that God in Christ takes the sins of the world upon himself, here that the powers of injustice are vanquished, here that death itself is overcome.

The cross, of course, is not the end. As Peter will later proclaim to his fellow Israelites, this "Jesus Christ of Nazareth, whom you crucified . . . God raised from the dead. . . . Salvation is found in no one else, for there is no other name under heaven given to mankind by which we must be saved" (Acts 4:10-12). After his resurrection, Jesus explains to his confused disciples that this was God's plan all along. If they understood their Scriptures (the Old Testament) aright, they would understand that "the Messiah will suffer and rise from the dead on the third day, and repentance for the forgiveness of sins will be preached in his name to all nations, beginning at Jerusalem" (Lk 24:46-47). This is the gospel. It is the good news about Jesus, about God in Christ entering his creation to make possible the forgiveness of sins for all peoples. It is the inauguration of the time of the Lord's favor, the age of the Spirit that brings salvation and newness of life in the present and points forward to a final renewal of all things (see Acts 3:18-21).

It is no surprise that the term *gospel* becomes in the New Testament the standard way of referring to what God has done through Christ and to the ongoing proclamation of this good news to the whole world. Here is what Paul highlights, for example, when he wants to remind the wayward believers in Corinth about the gospel that he had proclaimed to them (he may be merely reciting here a summary of the gospel that was already popular among the earliest followers of Jesus): Paul says that the gospel taught them "that Christ died for our sins according to the Scriptures, that he was buried, that he was raised on the third day according to the Scriptures, and that he appeared to Cephas, and then to the Twelve" (1 Cor 15:3-5). It is a very short summary. But notice how the good news here is above all a story that culminates in Jesus' death, burial and resurrection. The gospel is first and foremost the story of Jesus, the Messiah (Christ).

Paul says something very similar at the beginning of his letter to the Romans, where he describes the "gospel of God":

> the gospel he promised beforehand through his prophets in the Holy Scriptures regarding his Son, who as to his earthly life was a descendant of David, and who through the Spirit of holiness was appointed the Son of God in power by his resurrection from the dead: Jesus Christ our Lord. Through him we received grace and apostleship to call all the Gentiles to the obedience that comes from faith for his name's sake.... And you also are among those Gentiles who are called to belong to Jesus Christ. (Rom 1:2-6)

Note again the emphasis on how the promised good news is bound up with the story concerning Jesus, who is both the descendant of David—from whom the Messiah (Israel's anointed king) was expected to come—and the Son of God. Notice too how both here in Romans and in 1 Corinthians, Paul stresses that this gospel is "according to the Scriptures" (1 Cor 15:3, 4), and it was "promised beforehand through his prophets in the Holy Scriptures" (Rom 1:2). Paul's Scriptures, of course, are what we call the Old Testament. According to Paul and the other early believers in Jesus, the Old Testament told already of the promise of Christ. For Paul, the story of Jesus *is* the gospel, and yet this good news cannot be fully understood apart from the wider story of which it is a part—a story that began way back in Genesis. If we are to understand the significance of Jesus and the gospel, then, we need to know something about this wider context, this bigger story. As Christians seeking to be faithful to Jesus and the gospel, we need to go on reading the whole of the Bible, both the Old and the New Testaments. As we share our faith with others, we need to learn how to tell the story of Jesus—to proclaim this gospel—as a part of the whole story of God's purposes in creation and redemption.

It is interesting in this light to observe that when Paul addresses people who do not know anything about the God of the Bible, he finds it necessary to fill in the story a bit. In Luke's record of Paul's speech to the Athenians in Acts 17, for example, we find that Paul begins his proclamation of the gospel with an emphasis on God as sole Creator of the universe and as sovereign Ruler of all that he has created (Acts 17:24-26). Paul does not want his audience to misunderstand who God is; he is not just another local deity, another god who might be worshiped alongside others or added to the idols that Paul has observed all around him in Athens. The God of Israel is the one who made all that exists and who rules over all of his creation. But if God is wholly *transcendent* and apart from any created thing, he is also, Paul says, one who is at the same time near to every person and capable of being known by all (Acts 17:26-28). God is *immanent*, present to his creation and involved in it. For Paul, of course, this is displayed above all in Jesus, in the Son of God himself who has entered his creation and who reveals the character of God as both Creator and Redeemer. In the light of the revelation of such a God, Paul next moves (very quickly in Luke's retelling!) to call on his audience to respond. Paul stresses the need for all to turn away from idolatry and toward God in repentance (Acts 17:30); this sole Creator of all is also the Judge of all. There is, Paul says, a fixed day when "he will judge the world with justice by the man he has appointed," and "he has given proof of this to everyone by raising him from the dead" (Acts 17:31). The man Jesus, according to Paul, has been appointed by God to be the agent of judgment, for he is the resurrected Lord through whom God's justice and righteousness will finally be established.

Notice how Paul has moved in a few lines from God's sovereignty in creation and history, to humankind's failure to worship God, to the need for repentance in light of the certainty of future judgment by the resurrected Jesus. Luke's record of Paul's speech is obviously brief and undoubtedly selective, and the mixed response Paul gets from his audience suggests that there was more discussion yet to be had (Acts 17:32-34).

Nonetheless, we get a hint here of how some of the things that Paul could assume when addressing fellow Jews and Jewish Christians needed to be spelled out more fully when he spoke to those who knew little or nothing about the God of the Bible. This reminds us again—especially in our own mostly secular contexts—that the story of Jesus needs to be told as a part of the whole story of the Bible. And this is a story that begins with God's creation of the universe and ends with God's future judgment of sin, evil and injustice, and—as we will see—with new creation. It is not a story that begins with me; it is a story that begins and ends with God in Christ.

Paul stresses this point elsewhere when he talks about "the gospel that displays the glory of Christ, who is the image of God," and goes on to say, "For what we preach is not ourselves, but Jesus Christ as Lord" (2 Cor 4:4-5). The lordship of Jesus Christ, his rule over all of his creation (for he is also, according to Col 1:16, the one through whom God created the universe), stands at the center of Paul's gospel proclamation. So while we rightly emphasize the significance of this story for ourselves as individuals when we talk about how in Jesus we can have forgiveness of sins and be reconciled to God—something that Paul too has plenty to say about—we must remember that this gift of God's grace comes within the context of the Creator God's project to reconcile in Christ the entire world to himself so that he might reign over all (Col 1:19-20; 1 Cor 15:20-28). Our undeserved privilege as forgiven sinners is to escape the judgment that has been borne on our behalf by Jesus on the cross and to be accepted into the kingdom of God. As a result, we come under the lordship of Christ and find in him our true identity as God's children, freed from the power of sin and death. God's original purpose for us, that we might rule as his image bearers in his creation, is thus realized finally in Jesus, as Hebrews 2 explains. The power of death is broken (Heb 2:14-15), atonement for sins accomplished and new life made possible. The Messiah Jesus "has destroyed death and has brought life and immortality to light through the gospel" (2 Tim 1:10). As we will see in the following chapters, this inauguration of the rule of God in Christ has profound implications not just for us as individuals but for all of creation.

Let us return to our question again: What is the gospel? Here is one way it can be summarized: it is nothing less than the good news that in Jesus, the Son of God and Messiah of Israel, God has defeated the powers of sin and death and has inaugurated his restored rule over all of his creation. He has made provision through the incarnation, death, resurrection and ascension of Jesus for all people—Jews and Gentiles together—to receive forgiveness of sins and new life in the Spirit, enabling them to live forever as his children under the lordship of Christ. This good news is for the whole of the earth. It reveals the way in which God's purposes for all of creation are accomplished in Christ, the means by which a world racked by sin and corruption is renewed and restored to its Creator. It is good news for Israel, whose promised anointed King and Lord is come to restore his people and save them from their sins. And it is good news for Gentiles, who now find themselves joined to God's people in Christ, rescued from idolatry and sin and made joint inheritors of Israel's promises. The gospel is the story and the proclamation of what God has accomplished in Jesus, and it confronts all who hear this good news with the call to respond in faith. It invites everyone to join the community of God's children who have been reconciled to him in Christ and who now serve him as their King. It is also a gospel that drives action. Paul calls this action the "obedience that accompanies your confession of the gospel of Christ" (2 Cor 9:13).

This "good news," with its proclamation of Jesus as Lord of all, is of such cosmic sweep that it necessarily challenges any and all other claimants upon our worship. It rules out any others who would demand our absolute allegiance or tempt us to give ourselves over to their service instead of to the service of God in Christ. When the gospel was proclaimed in the first century, it necessarily overturned literal idolatry of all sorts and also challenged any temptation to buy into the propaganda of the Roman Empire, with its own "gospel" of peace. This was a "peace" bought with war and violence, and its bearer was not the God of the Bible but the emperor. The emperor was thus sometimes hailed as a "son of God" himself, and the birth of Augustus could even be called "gospel," good news for the world.[11] But if Jesus is Lord of all, no one and nothing else can be. For faithful Jews who turned to Christ, even God's gift of the Old Testament law and the identity it conferred on them as his people now had to be seen differently in the light of the coming of the Messiah Jesus. Through this same Jesus both they and newly converted Gentiles were saved, and he became the one in whom their identity and unity as God's people was now to be found.

Today, the same gospel goes on challenging us, forcing us to question where our true allegiance lies. It asks us where we look for our salvation and our identity. It demands that we not buy into the competing claims of our own societies and cultures if we are to honor Christ alone as Lord. For those of us in the so-called developed world, the gospel must challenge any attempt to find our salvation, identity, self-worth and security in consumerism, in sex, in accumulation of wealth, in nationalism or in our worship of individual freedom to do whatever we want—a freedom that so often is borrowed at the expense of others. It also challenges us to stop placing our trust in spirituality, in religiosity or in popular secular ideologies of self-help. And since it is God alone who saves, the Christian gospel challenges any notion that we are the ultimate saviors of the planet, that we are the measure of all things, that in the end it is all up to us. In the end, as in the beginning, the gospel tells us, it is up to God in Christ.

Two Dangers and How to Avoid Them

There are two apparently opposite dangers that we must avoid if we are to be faithful to what the Bible teaches about the gospel and the future. The first danger is to assume that biblical hope is of the "pie in the sky, by and by" variety that limits concern with this world and shuts down engagement with the difficulties and challenges inherent to living here and today. Such an assumption is often linked to views of the gospel as something all about me, as something that speaks only to my own existential crisis or answers *only* to my individual plight as a sinner before a holy God. We have already begun to see that though this is a popular caricature of Christian belief, it does not reflect the biblical perspective of the gospel taken as a whole. It represents an attempt to reduce the good news about Jesus to a story about me and God, rather than a story about God in Christ that is for and about the whole world.

The opposite error is equally seductive, however. This is to give up on biblical hope in the face of its despisers, to suppress the reality of our own rebellion against God and our need for his mercy, and to assume that Christian faith can be reduced to a pattern of living in the present that gives no consideration to the future. This represents an attempt to excise the central message of the gospel as it addresses us as individuals while trying to keep what we deem to be the moral or practical benefits of a "Christian worldview." Ironically, the failure of this approach is again precisely in its focus on ourselves rather than on God in Christ. It presumes that we are capable of solving the fundamental problems of the world by ourselves, that we have no need of a Savior but only of a moral exemplar. The biblical perspective is far bleaker with regard to our human nature, far more honest about our sinfulness and our brokenness. But it is also far more radical in its promise of transformation and renewal, a transformation that begins now. Its vision is further reaching and all-encompassing in its hope for a new creation toward which we may indeed work but which finally is given only by the grace of God himself.

We are convinced that pursuing either one of these options in our thinking about the future will diminish our effectiveness in dealing with the very real challenges that our world faces. The "pie in the sky, by and by" version of otherworldly hope gives us permission to ignore the present world and its suffering. It presents us with dreams conveniently unrelated to our everyday life. The "day by day and in every way things are getting better and better" progressive optimism in human ingenuity to solve all of our problems dispenses with Christian hope altogether. It represents the perpetuation of the same ideology that is responsible for most of the problems that we face in the first place. It fails to reckon with our limitations and our sinfulness, rejecting what the Bible would tell us about the reality of our plight apart from God's grace in Christ. If we place all of our faith in ourselves, if we fail to remember that we are limited and flawed creatures, if we imbue our own plans for the future with an ultimacy that they cannot bear, we will end only in frustration, disappointment and despair.

Above all, pursuing either of these options will lead us astray from the gospel proclaimed to us in Scripture. The way to avoid both of these dangers in the context of our discussions about eschatology is always to center our thinking about the future on the resurrection of Jesus. It is here that we must begin, with what God through the Holy Spirit has done in Christ. Jesus' resurrection is what founds and shapes Christian hope and gives it its content and its meaning. The resurrection confirms and displays God's purposes for all of his creation, and it reveals to us that the future is not simply a time or a place: the future is a *person*. The future is God-in-Christ whose welcoming arms embrace all that he has made and draw us forward to our promised rest when we will at last see him face to face. In the chapters that follow, we spend a lot of time thinking more specifically about the place of the *earth* within biblical eschatology; but we will do well to remember that Jesus' cross and resurrection must shape all of our discussions not only about our own future as individuals but about the future of life on earth too.

Looking Ahead

We observed earlier that Paul tells us that "the greatest . . . is love," that love must be always at the center of all we do. As we finish our assessment in this chapter of the place of hope in the Christian life and its potential to shape our environmental ethos, it is worth returning to 1 Corinthians 13 to recall that Paul also affirms here that faith and hope remain indispensable in this in-between time of imperfect knowledge and limited sight, in this age of seeing only dimly as in a dark mirror (1 Cor 13:12). Christian ethics find their wellspring in the life-giving power of Christ's resurrection. They are necessarily oriented toward the kingdom of God proclaimed by Christ, a kingdom that has broken into the present and yet remains only partially realized. As Jesus taught his disciples to pray, Christians still yearn for the day when God's will is fully done "on earth as in heaven." In the chapters that follow, we ask what that might look like.

5

Bringing New Testament Hope Down to Earth

Heaven is a place on earth.

BELINDA CARLISLE

I saw the Holy City, the new Jerusalem, coming down out of heaven from God.

REVELATION 21:2

◆

MINDS IN HEAVEN

There is an old stereotype of Christians that they are "too heavenly minded to be of any earthly good." We have probably met a few who fit the part. Heads in the clouds and focused on "spiritual" things, they fail to be of much practical use to anyone and have little interest in what is going on in the wider world. They appear detached from the mundane things that concern us ordinary mortals. It might even be that we fit this caricature sometimes, or perhaps wish we did. Especially in the midst of trouble, sorrow, temptation or despair, we can find ourselves—in the words of the twentieth-century hymn "Turn Your Eyes upon Jesus"—longing for "the things of earth [to] grow strangely dim."[1]

If, however, by "heavenly minded" we mean having something of God's priorities in view, adopting his perspective on life, is it possible for someone who is truly heavenly minded to be of no earthly good? When we turn our eyes upon the Jesus who is revealed to us in the New Testament, we surely do not find that the things of earth grow dim, dissolve or blur into a haze. Rather, as indistinguishable shapes and shadows resolve themselves into rocks and trees and hills when the sun rises and dissolves the mist of an autumn morning, we discover in the light of Christ that all of reality becomes clear and sharp. Earthly things are seen finally for what they really are, even as their mastery over us and their power to lead us away from God is diminished (which, to be fair, is probably the point of the otherwise lovely hymn).

What we mean by "heaven" and how we conceive of our Christian hope has important implications for what it means to be heavenly minded people on earth. Getting this right is crucial if we desire to orient ourselves toward God and God's future while living in a world facing the challenges summarized in chapters 2 and 3. In this chapter we consider what the New Testament has to say about heaven, but above all we will give our attention to Paul's stunning portrayal of the future of *earth* in Romans 8.

HEAVEN AND EARTH

The vision of the future held out to God's people in the Old Testament is earthly, material and this-worldly. Just as the land and all of creation suffer as a result of the rebellion of God's people against him, so will this creation be renewed when God acts to save and restore his people. But what about the New Testament? It is often suggested that the New Testament shifts its focus entirely to "heaven" so that any earthly hope is spiritualized, relegated to a brief interim period (such as the millennium) or done away with entirely. In this scenario, it is difficult to see that there can be any genuine future for life on earth; human beings, perhaps, will finally escape earth for heaven and leave the rest of creation behind.

There is no doubt that the New Testament has a lot to say about heaven—and not merely heaven in the sense of the sky above the earth. Just as the Old Testament portrays God's throne in the heavens (e.g., Ex 24:9-10; 1 Kings 22:19; Ps 11:4; Is 66:1), so too does the New Testament assume that heaven is supremely God's realm (e.g., Mt 5:34; Heb 8:1). It is depicted as a place apart from the earth and its fleeting, corruptible treasures (Mt 6:19-20), a place where the lasting rewards of Jesus' persecuted disciples are kept safe (Mt 5:12; cf. 1 Pet 1:4). John is given the privilege of ascending to heaven and seeing a vision of God on his throne in the heavenly throne room (Rev 4:1), and Paul hints that he too had such an experience (2 Cor 12:2-4). Jesus talks of rooms prepared for his followers in his father's house (Jn 14:1-4), and it is difficult to conceive of this house as located anywhere other than heaven. Paul similarly describes a "building from God, an eternal house in heaven," although his focus is especially on this building as the "heavenly dwelling" of the resurrection body, when "what is mortal" is "swallowed up by life" (2 Cor 5:1-4). Paul elsewhere tells the Philippians that their citizenship is now in heaven (Phil 3:20), the place from which they await the coming of their Savior. Likewise he encourages the Ephesians that in Christ they have already been raised to new life and are seated in the "heavenly realms" (Eph 2:6). The author of Hebrews—who comes nearest to using the language of "heaven" in the way it is popularly used today—talks of God's faithful people as sojourners on earth, longing for a homeland, a "better" place than the one they left, a "heavenly one" (Heb 11:13-16).

Heaven in the New Testament may refer either to the dwelling place of God, the place of his throne where his will is done and his reign is uncontested, or to the place where the rewards of the righteous are kept safe. This is the place to which his people should properly orient themselves as citizens and the place of those things for which they long: the heavenly city, a new Jerusalem, resurrection life and, above all, their Savior Jesus Christ, who one day will return from heaven. In this sense, Christians have every reason to yearn for heaven, to fix their eyes on what is unseen—on God's eternal glory—in the face of suffering and persecution (see 2 Cor 4:18). As it did for Paul, the sure and certain hope that our future is kept safe with God in Christ can enable perseverance and faithfulness in all that God calls us to in this life. The language of "heaven" as it is used by the New Testament writers emphasizes that Christ himself, as well as our new identity in him and our future hope, is not of human or earthly origin. It warns against the idolatry of seeking the source of our life, identity and hope anywhere else.

It is important to recognize, however, that the New Testament emphasis on heaven as the source of the hope of God's people does not mean that the New Testament writers have given up on the earth. At best this would be an argument from silence, based only on those texts that emphasize our "heavenly" hope without specifying one way or another what this means for the earth. In these cases we could as well argue, in keeping with the Old Testament perspective, that the rest of creation itself will be renewed when God's people come into their heavenly inheritance and are made to be those whom God intends them to be. And as it turns out, this is precisely what we do discover when we look at those texts that actually specify the place of the earth within Christian hope.

The New Testament nowhere suggests that heaven is our final dwelling place. In fact, what we sometimes mean by *heaven*—a place in the sky where we hope to go when we die and live forever away from this earth—turns out to be misleading. The New Testament says less about people going to heaven than it does about heaven being realized on earth. It teaches us to pray for God's will to be done "on earth as it is in heaven" and promises us that God will come to dwell with his people at the renewal of all things. Notice how even when Paul talks about our citizenship in heaven in Philippians 3:20, he does not therefore conclude that heaven is the place to which its citizens might hope to go one day. Rather, by analogy with the way that Roman citizens living in Philippi conceived of their relationship to the Roman emperor, who might one day come to their city from Rome, heaven is the place from which heavenly citizens await the coming of their Lord and Savior, Jesus Christ. While we can with Paul be confident that at death we "depart" to "be with Christ" (Phil 1:23), the Christian's ultimate hope according to Paul is for Christ to come to *earth* (Phil 3:20). In Revelation we learn that even the heavenly city—the new Jerusalem—comes down out of heaven to earth in the new creation (Rev 21:2). At that time the distinction between earth and heaven will no longer matter because God's glory and presence will be displayed fully throughout the new heaven and new earth.[2]

In the remainder of this chapter we will focus on Romans 8, a passage where Paul tells us more clearly than anywhere else just what the relationship between the present creation and the future promised us in Christ looks like. This is not to say that Romans 8:18-24 is quite as easy a text to interpret as we might like. But with a bit of exegetical work and attention to the wider biblical context, what Paul affirms about the future of creation in this passage is clear enough to dispel any notion that biblical hope is thoroughly other-worldly or only "heavenly." This is why we begin our detailed treatment of the New Testament with this text: the power of Paul's vision of the future as it is described in Romans 8:18-24 makes this text an appropriate foundation for the construction (or, we might more boldly claim, the reconstruction) of a biblical model of Christian hope. To switch metaphors, this text is the clearest lens through which to read what the rest of the New Testament has to say about the future of earth. What in other passages is left unspecified is spelled out in the cosmic vision of Romans 8.[3]

THE GLORY TO COME

Paul makes two bold and startling claims at the center of his letter to the Christians in Rome. They are claims that give us some insight into the power of the future to shape Paul's perspective on the present. First, in Romans 8:18 Paul says, "I consider that our present sufferings are not worth comparing with the glory that will be revealed in us." Then, after he has described something of what this glory looks like, he makes the point even more strongly in verse 28. Here he implies that the sufferings of the present time might even have a positive role to play in leading to the incomparable glory of the future: "In all things God works for the good of those who love him" (Rom 8:28). Paul is convinced that everything, even suffering and loss, pales nearly to insignificance when seen in the light of the glory that is to come. All of the pain and sorrow, sickness and death that permeates this world—against which all of us struggle and on account of which we are sometimes tempted to give in to despair—all of this is finally not worth comparing, Paul says, to the glory that will be revealed at the future resurrection (cf. Rom 8:23).

How can Paul, someone intimately acquainted with pain and loss himself (see 2 Cor 11:24-28), be so confident about the future and so prepared in its light to discount the weight of present suffering? The answer is found in the wider context of Romans 8. Paul says that he and all of those who are in Christ have been adopted by God to be his children. This is a present reality to which the Spirit testifies (Rom 8:14-17). As God's adopted children, all believers have been guaranteed a precious inheritance: they are now heirs with Christ of what God has and of what God gives (v. 17). Above all, God has predestined his adopted children "to be conformed to the image of his Son," Jesus (Rom 8:29). This is why Paul can say—even in the face of suffering and death—that "in all things God works for the good of those who love him" (Rom 8:28). Paul's paradigm is the life, suffering, death and resurrection of Christ. In Jesus, Paul finds the pattern for his own life and for the lives of all of his fellow believers.

Paul once knew Jesus only at a distance, as a man who had suffered and died on a Roman cross, despised by his people and apparently rejected and cursed by God. But on the road to Damascus Paul encountered this same crucified Jesus as the risen and glorified Lord (see Acts 9:1-19; Gal 1:11-24). Paul now sees in Christ the supreme revelation of God's glory; in Christ God has entered his creation, taken the sins of the world upon himself and "condemned sin in the flesh" (Rom 8:3). This is why Paul can announce boldly that there is no longer any condemnation for those who are united with Christ (Rom 8:1). God's victory over the powers of sin and death has been definitively proclaimed in Christ's resurrection and glorification. And just as Christ was raised from the dead, so too Paul knows that one day "all will be made alive" (1 Cor 15:22). It is at this final redemption that the full identity of God's children will be revealed. It is for this day that they and all of creation long, the day when they will share in the glory of God as revealed in Christ (Rom 8:19-23).

Paul's entire life now is driven by his desire to imitate Christ, and he expects the same of all of those who trust in Christ as Lord. To be adopted as a daughter or a son of God is to become an imitator and follower of *the* Son of God, Jesus. To be a child of God is moreover to share in the hope of resurrection promised in Jesus' resurrection. And this hope for resurrection, for the "redemption of our bodies" (Rom 8:23), is actually something for which all of creation yearns (Rom 8:19-22).

A Groaning Creation

Why should the nonhuman creation care about the resurrection of the children of God? It is because, according to Paul, the future of the entire creation is bound up with the future of God's children. For evidence of this, look closely at how creation is portrayed in Romans 8:18-25:

> I consider that our present sufferings are not worth comparing with the glory that will be revealed in us. For the creation waits in eager expectation for the children of God to be revealed. For the creation was subjected to frustration, not by its own choice, but by the will of the one who subjected it, in hope that the creation itself will be liberated from its bondage to decay and brought into the freedom and glory of the children of God.
>
> We know that the whole creation has been groaning as in the pains of childbirth right up to the present time. Not only so, but we ourselves, who have the firstfruits of the Spirit, groan inwardly as we wait eagerly for our adoption to sonship, the redemption of our bodies. For in this hope we were saved. But hope that is seen is no hope at all. Who hopes for what they already have? But if we hope for what we do not yet have, we wait for it patiently.

The first thing we learn about creation in this passage is that creation "waits in eager expectation" for the same thing for which we wait, "for the children of God to be revealed" (v. 19). In light of the wider context we have just discussed above, it seems odd that this is something for which we or creation should still have to wait. After all, Paul has made it clear in verses 14-17 that he considers those led by God's Spirit *already* to be children of God. The explanation for this apparent contradiction is found in verse 23. Here Paul explains that this revelation of the children of God represents something more: it represents the future result of our adoption as God's children, the full realization of our status when we finally enter into our inheritance and experience the resurrection, the "redemption of our bodies."

As is so often the case in Paul's writings, there is both an *already* and a *not yet* aspect to our identity as God's children. We are *already* assured of our status before God in Christ and are *already* expected to live as God's children, as those who are led by the Spirit of God. But we do *not yet* see or fully experience the glory that God has in store for us. Our lives—if they are anything like Paul's—are marked by suffering and death, and the world in which we live remains a world shot through by brokenness and grief. It is in the midst of this tension between the old age and the new that we yearn for the final redemption and new creation.

Returning, then, to verse 19, we have observed that the creation itself shares with us in this longing. But what does Paul mean by "creation" here? Readers of the King James Version might notice that it renders the Greek word *ktisis* here as "creature" (although it uses "creation" for the very same Greek word in v. 22). Nonetheless, nearly all interpreters today recognize that, based on how this word is used elsewhere,[4] Paul must be referring in all of these verses to the entire nonhuman creation. The word *ktisis* can often be used for human and nonhuman creation together, but since in this passage Paul distinguishes between the "creation" and the "children of God," it is best to interpret "creation" here as referring especially to all of the nonhuman created order. We sometimes use the word *nature* with this sense. But since *nature* has so many different meanings and possible connotations (and fails to make explicit that it depends on a Creator), it is better to use *creation* to designate the entire living and nonliving world that is created and sustained by God.

It is this entire creation, then, that is waiting expectantly—standing on tiptoe (as some interpreters have rendered the Greek expression here)—for the future revealing of God's children. Why is it so eager to see this future realized? Paul says that it is because everything is not right with creation now. It has been "subjected to frustration" and longs to be released from its "bondage to decay."

Paul assumes here, as he so often does, that his readers know something about the wider biblical story. He is taking us back to Genesis, where humankind's rebellion against God tragically disrupted the relationships between God, human beings and creation itself. Adam and Eve were placed in Eden to do God's work, to bring about his purposes for creation as his image bearers, and so their disobedience and rejection of God's command could only mean the frustration of God's intentions for all of creation. God subjected creation to humankind, his royal image bearers; and when these rulers within creation become corrupt, self-seeking, God-denying and a law unto themselves, all of creation suffers as a result. In Genesis 3:17 this is seen already in the cursing of the ground itself as a result of Adam's disobedience. In the immediate context of Genesis, this "curse" is experienced as a brokenness in the relationship between Adam and the earth, a brokenness that manifests itself in the difficulty that Adam will now have in getting food from the ground. Paul, however, like the Old Testament prophets before him, goes further in describing how creation's subjection to now-fallen humanity means that the entire creation is subjected to ongoing frustration, finding itself in "bondage to decay" (Rom 8:21), enslaved to ruin. God subjected creation to human beings on the basis of the hope that all creation would one day share in the glorious freedom of God's children (Rom 8:20-21); but until that day, creation's subjection to fallen humanity is experienced as a subjection to futility and ruin.

What does this look like? What does it mean for creation to be "groaning" and in "bondage to decay"? Although some people today might talk about the suffering of "Mother Earth" or "Gaia," it is otherwise rare to hear the entirety of nonhuman creation personified in the way that Paul does here. Christians in particular tend to be wary of such language, concerned as we are to disassociate ourselves from the apparent idolatry that sometimes attends this way of talking about the earth. Most scientists are also rather strict about avoiding anthropomorphic language to describe nature. Paul has no such qualms. He is able to draw on a rich biblical tradition of letting nonhuman creation have its own voice, a voice that is heard praising God throughout the Psalms, bearing witness to the covenant between God and his people in the Prophets and, as here in Romans 8, crying out—groaning even—when creation suffers the results of humankind's corruption.

Paul is in fact echoing the language of Isaiah 24–27, a passage that he uses in his extended defense of the hope of resurrection in an earlier letter, in 1 Corinthians 15. He alludes to the same passage again in his description of life after death in 2 Corinthians 5:4. Just as Paul does in Romans 8, Isaiah 24–27 emphasizes both the present devastating effects of human sinfulness for a mourning earth and also the cosmic extent of the judgment and new creation to come. At that time the Lord will swallow death for ever and will provide new life to his fainting people, a people whose own efforts cannot, despite their groaning as in labor pangs, bring forth the deliverance for which they long and hope.[5]

At the beginning of this section of Isaiah, in chapter 24, we read about the terrible consequences for the earth that have resulted from its subjection to those who live on it, and about the cosmic breadth of God's judgment:

> The earth dries up and withers,
> > the world languishes and withers,
> > the heavens languish with the earth.
> The earth is defiled by its people;
> > they have disobeyed the laws,
> violated the statutes
> > and broken the everlasting covenant. (Is 24:4-5)

Notice how the cause of the earth's "defilement" or "pollution" is spelled out more fully in Isaiah 24 than it is in Paul's brief description in Romans 8. According to Isaiah, it is due to the actions of the earth's human inhabitants, those who have violated God's law and broken the eternal covenant. That covenant is the bond established between God, his people and his creation that is intended to bring *shalom*—peace, wholeness, healthfulness and fruitfulness. We have already seen how Paul's description of creation's groaning in bondage to ruin in Romans 8 deliberately echoes the story of Adam and Eve in the Garden of Eden and the results of their rejection of God's purposes. Here in Isaiah 24 we see the way in which this tragic drama continues to be played out in the ongoing relationship between human beings, God and creation. The sin of Adam and Eve and the resulting curse is not simply a one-off event; it is the paradigm for how the ongoing rejection of God by the children of Adam and Eve continues to bring about brokenness in the relationship between God and his people. That brokenness extends also to the relationships of humans with each other and to their relationship with the rest of creation.

The drying up and withering of the earth in Isaiah 24 is set in the context of a description of drought and ruin, the failure of crops and the desolation of cities. For the people of Israel, these were very real threats to their well-being on the land and to their existence as a nation. Yet Isaiah 24 goes further in linking such localized disasters to the plight of the entire creation: "the world languishes and withers, the heavens languish with the earth" (v. 4). The prophet Hosea, as we observed in chapter 2, paints a similarly cosmic scene of the effects of human injustice on the earth. Describing a time when "there is no faithfulness, no love, no acknowledgment of God in the land" (Hos 4:1), Hosea says that as a result "bloodshed follows bloodshed" (Hos 4:2) and "the land dries up [or "mourns"], and all who live in it waste away; the beasts of the field, the birds in the sky and the fish in the sea are swept away" (Hos 4:3). Behind both Isaiah 24 and Hosea 4 no doubt lie very real, localized experiences of ruin and devastation. Yet in the light of biblical tradition stretching back to Genesis, Isaiah and Hosea perceive these as but particular instantiations of the plight of the whole of creation. It is a plight that has resulted from creation's subjection to sinful and rebellious humankind; and it is a plight that calls ultimately for a divine act of cosmic renewal and restoration.

This returns us to the question of just what Paul means when he describes creation as subject to "frustration," in "bondage to decay" and "groaning as in the pains of childbirth." Since we have seen that Paul's insight into the state of creation is rooted in the wider biblical tradition, especially Genesis 3 and Isaiah 24–27, his conclusion is likely to be first and foremost a theological rather than a merely empirical one. In other words, Paul's claim is not based simply on his observation of nature but on his interpretation of the state of creation when viewed through the lens of Scripture. The frustration of creation in the context of Romans 8, for example, is seen above all in its failure to reach the end or goal for which it is intended: to bring glory to God. Human beings, who ought to perceive in creation a reflection of the glory of the Creator God and worship him alone, have instead rejected God and turned to the idolatrous worship of creation itself (Rom 1:21-23).

At a very basic level, human beings have made and worshiped actual idols constructed out of God's created matter. Both pagans and God's people (who should have known better) have continually made such idols, as both the Old and New Testaments report over and over again (e.g., Ex 32:1-8;

Num 33:52; Deut 29:17; Ezek 20:7-8; 1 Cor 8:4; 12:2; 1 Thess 1:9). The New Testament extends the meaning of idolatry even further, however, by equating it with such things as greed and covetousness (Eph 5:5; Col 3:5), perceiving that idolatry is the worship of *anything* that is not God. This perhaps speaks most directly to those of us in Western societies and high-income countries, where the more abstract idols of wealth, fame, sex and power are often those that attract us and consume our lives. We do not need to construct idols from wood or metal—though it is certainly easy to waste much of our lives in the pursuit of the newest electronic gadgets, the best performance cars, the most fashionable clothing or the latest recreational equipment. In the light of what Paul says in Romans, our consumer societies actually hinder the material creation itself from fully glorifying God because we have hijacked it for selfish (and therefore sinful, hence empty) ends.

Paul still affirms that creation reflects God's glory, and—as the Psalms and the book of Revelation make clear—it may yet be seen as praising God in and of itself. But creation cannot finally reach the end for which God intends it when those to whom it has been subjected—namely humans, the very image bearers of God—not only fail to join in the worship of the Creator but even twist God's purposes by making created things themselves into idols.

The futility to which creation is subject leads to its "groaning as in the pains of childbirth" as it longs to be freed from its "bondage to decay." This "bondage to decay" refers to the way in which creation, by being subject to fallen human beings, has become enslaved to "corruption" and "ruin" (as we might translate the Greek word *phthora*, which lies behind the NIV's "decay"). As Paul's use of Isaiah 24–27 suggests, this highlights the way in which humankind's ongoing rebellion against God leads to the suffering of creation itself. Creation does not yet experience the freedom that God intends for it. This, again, is first and foremost a theological insight into the way in which an otherwise good creation experiences brokenness in its relationship to humankind, and the way in which this brokenness results in ruination and leads to creation's groaning. But the fact that Paul uses an analogy of the pains of childbirth draws attention again to the long-term perspective of the hope for the future of new birth, of the new beginning that will come with the new creation. Labor pains are certainly agonizing and all-consuming at the time. Yet they are quickly forgotten and relegated to the past in the joy of birth and the arrival of new life. It is an apt and encouraging metaphor for us as we endure the pain and brokenness of this world.

In Old Testament times Isaiah and Hosea could discern in the devastation of cities and land, in drought and in crop failure, signs of a fundamental cosmic disorder. No doubt Paul too found in similar natural disasters of his day confirmation of the ways in which all of creation continues to groan and long for its freedom. First-century residents of the Roman Empire were of course as well acquainted as people of any era with natural disasters, many of these caused by forces beyond their control, others more obviously linked—even then—to human overexploitation of the local environment.[6] When we read Romans 8 today, we cannot help but see—and indeed ought to see—creation's groaning reflected in our current ecological crises (see, e.g., figure 5.1), especially now that the truly global consequences of our actions for the rest of creation have become so evident.

Figure 5.1. A groaning creation? Fort McMurray, Alberta, Canada: an aerial view of tar sand mines and tailing ponds. (Photo by Veronique de Viguerie/Getty Images)

If the biblical picture of humankind's role within creation once appeared naive to some for the way in which it assigns such profound responsibility for the earth to one species, it no longer appears so—not in an age when human beings are having such widespread effects on the earth that scientists have begun to call it the "Anthropocene," or "Age of Man."[7] The need for us to take seriously our responsibility for creation has never been greater, and the potential consequences of the failure to exercise our responsibility well have never been so cataclysmic.

A Future for Creation

Why did God subject creation to humankind if it has led to such disastrous consequences for the earth? According to Romans 8:20, this is because the present suffering of creation is not the end of the story. Its subjection was done "in hope," on the basis of God's intention that the rest of creation should share not only in the results of the fall of human beings but also in their redemption. Creation will one day be freed from its slavery to ruin and enter into the glory that accompanies the redemption of the children of God.

We are reminded again of the way in which creation's fate has been linked to humankind since Genesis 1–3, where human beings are appointed to rule within creation as God's image bearers (Gen 1:26-28) and Adam is told to "work . . . and take care of" (Gen 2:15)—or, as we might translate it, "serve and protect"—the land where he is placed. When read in the light of the New Testament, the Genesis story hints at a direction for creation, a purpose that God intends to fulfill through his appointed human creatures. In Romans 8, Paul observes that this purpose appears thwarted when the rule of human beings in creation leads, on account of their corruption and fall, to creation's subjection to corruption. But God's ultimate purposes are not thwarted. In Christ, the new Adam and the true image of God, creation's glorious future has been secured; it too will share in the freedom that attends the redemption of the children of God.

Paul does not give us any details about what this freedom looks like apart from suggesting that it means the undoing of the present futility of creation. Creation will fulfill perfectly and completely the purpose for which God intends it. It will mean its liberation from decay, corruption and ruin, since the children of God will finally fulfill the role in Christ for which they were intended. The brokenness that marks life in this age, brokenness in the relationship between human beings and God and between human beings and creation, is healed in Christ. In Christ, God's covenantal purpose of *shalom*, of peace, wholeness and flourishing life, is brought to completion. A more extended description of what this looks like in the new heavens and new earth awaits us in Revelation 21–22, although even there we will see that the glorious future of creation promised in Scripture so exceeds our imagination that only symbol and metaphor can begin to describe it. For now it is enough—and indeed it is absolutely vital—to observe that Romans 8 makes it clear that this very same creation that is groaning now has a future in God's ultimate purposes. Whatever else we might conclude about the details of our Christian hope, we must affirm in the light of Romans 8 that *this creation*, this very earth, will not be left behind.

What Difference Does Creation's Future Make Now?

We began our study of Romans 8 by observing how Paul's profound hope in the future enabled him to discount the weight of present suffering and to follow in the way of Christ, in the way of the cross. Given that Paul expects all of those who are in Christ to adopt the same attitude and to imitate him as he imitates Christ (1 Cor 4:16; Phil 2:1-18), we have to ask whether Paul's portrayal of creation's future in Romans 8 affects how we follow Christ today. Does it make any difference to how we live now when we recognize that the earth's future is inextricably linked to our future as God's children?

As it turns out, it ought to make a tremendous difference. If we take seriously what Paul says about our present and our future, it will transform how we relate to the world today and how we approach many of the challenges facing life on earth. Consider first of all the value that creation has for God and the significance of that value in our decisions about how we treat the earth. In 1 Corinthians, Paul finds it necessary to challenge the Corinthians to take seriously how they treat their own bodies. Some of the people in the church at Corinth apparently thought it did not matter what they did with their physical bodies; perhaps they thought that their bodies did not matter now that they had become so "spiritual." Whatever the reason for their muddled thinking, Paul's message to them—in a book that culminates with an entire chapter on the resurrection (1 Cor 15)—is that they were "bought at a price" and so must "honor God" with their bodies (1 Cor 6:20). Christ's blood was not shed to give them a means of spiritual escape from this world. Rather, they were redeemed as whole persons. Their bodies were themselves purchased by Christ's blood and made able to share in the hope of resurrection. Therefore, the way they treat their bodies matters to God. In Romans 8, we have seen that Paul links the hope of all of creation to this hope for the "redemption of our bodies." This link of creation's fate to ours suggests that the entire material creation was "bought at a price"; its future too has been secured only through the blood of Christ. What Paul tells the Corinthians about treating their physical bodies as valuable to God therefore can also be seen to apply to the entire creation. The way we treat creation matters to God.

Second, consider again what Paul says about our status as children of God. Although we do *not yet* see the full revelation of our status, which will occur only at the resurrection, we are *already* adopted as God's children and are *already* expected to live in the light of that: we are to live as those who are led by the Spirit (Rom 8:12-17). If the entire creation longs to see us become who we are—to have our status as God's children revealed—our lives even now ought to begin to be orientated toward creation in a way that is in keeping with God's ultimate purposes. They ought to be lived in a way that is in keeping with creation's eventual freedom from the futility and ruin that it suffers at our hands. Paul elsewhere says that when anyone is in Christ, there is "new creation" (2 Cor 5:17). God's future new creation breaks into the present in Christ and in all of those who through Christ by the power of the Spirit become God's children. We are called to live now as signposts pointing to the new creation and, more than that, even as instantiations of the new creation here and now. We are called in all of our relationships to imitate the humility and love of Christ, who laid down his very life for others. Through Christ, God reconciles to himself all things (Col 1:20). As we find ourselves reconciled to God as a part of his cosmic peace-making, so too must we be reconciled with each other and with all of creation.

When we take seriously the cosmic scope of Christian hope and our call to live as God's children now, we find that the scope of our love and of our ethics extends even beyond our fellow human creatures to embrace all of God's creation. He values it; we must value it too. Our casual selfishness in how we use the earth's resources, in how we treat our global neighbors and in how we treat creation itself is revealed for what it is: an affront to God, an abrogation of the responsibility he has given us and a rejection of our identity as his children in Christ.

Conclusions

Paul's perspective in Romans 8 is sometimes labeled "anthropocentric": in his vision, creation's future is inextricably bound up with that of humankind. Science, of course, would tell us otherwise. If human beings disappeared tomorrow, the world would not end; the universe would go on expanding, the galaxies would still glitter in the vast expanse of space, our own earth would continue in its track around our sun. Though the effects of our presence would linger on the planet for millennia, life would carry on—and might in fact seem quite a lot better for many species. This is all until the sun one day expands in its dying and consumes the earth and, many more billions of years hence, the entire universe either collapses in on itself or continues expanding into cold nothingness. But just as what we know with scientific certainty about the finality of death in this life in no way undermines our faith in God to bring new life in the resurrection, so what we know through science about the fate of the universe and the future of our own earth does not undermine what Scripture reveals about the new creation. If God is able to give new life to our mortal bodies in a way that we cannot possibly comprehend and yet has been done already in the resurrection of Jesus, he is more than able to give new life to an entire groaning creation—to make *everything* new (Rev 21:5). Paul's astonishing assumption that the fate of a vast and apparently indifferent universe is linked to ours is rooted in the purposes of God from the beginning that all of creation should come to share in the glorious freedom of the children of God. It is rooted too in the recognition that God is not a God who sets up the mechanism of creation and then abandons it. He is not an absentee landlord. As we will see in the next chapter, God is a God who is involved in his creation and comes to his creation in judgment and salvation.

On a cosmic scale, this can be grasped only through faith in the God who raised Jesus from the dead. But on a more human, everyday scale—where our lives, and the lives of our children, are actually lived out—the biblical vision comports frighteningly well with what we have seen science is coming to realize about humanity's outsize role on the earth today. The biblical vision challenges us to take up our responsibility as God's children, to reflect God's love for his creation with wisdom and humility and to live even now as those who belong to the new creation.

Romans 8 tells us that this very same creation that today groans under the weight of human sin will be restored and liberated from its bondage to ruin at the revealing of God's children in the resurrection and new creation. This text confirms for us the value of nonhuman creation before God, and it reminds us of the ways in which our rebellion against God has led to its current plight. Above all, Romans 8 holds out for us the promise of a glorious future in the light of which our suffering now will seem light and momentary. This is a future given by God's grace and yet toward which we are called to orient ourselves even now as God's children. In its affirmation of creation's value, its linking of creation's fate to ours and its certain hope for the future, we discover in Romans 8 the foundations for a radical Christian environmental ethos.

6

Cosmic Catastrophe?

Some say the world will end in fire,
Some say in ice.
From what I've tasted of desire
I hold with those who favor fire.

ROBERT FROST, "FIRE AND ICE"

◆

PETER VERSUS PAUL?

Predictions of impending environmental catastrophe are popular among green activists and even some scientists. Such prognostications may not be surprising given the unprecedented situation in which we and the planet find ourselves. Christianity, however, has always had its own apocalyptic visions of the future. Whereas secular prophecies of doom are based on extrapolation of current trends into a future otherwise much like the present, Christian expectations are rooted in hope for a future that is dramatically different from the present.[1] Such faith is founded in the biblical promise that God will ultimately intervene to set things right in the world—to judge evil and to save a broken creation. Of the biblical texts that portray this future coming of God and his final judgment and new creation, probably none, apart from the book of Revelation, has been as influential as 2 Peter 3. This passage describes a cosmic catastrophe beside which even the direst effects of global climate change or ecosystem collapse appear tame by comparison.

We discovered in the last chapter that Paul makes it unmistakably clear in Romans 8 that the same creation in which we now live will participate in the new creation to come, that the Christian's longing for resurrection is matched by creation's longing for its own future liberation. But for many readers, 2 Peter 3 seems to challenge us with a quite different picture and even call into question Paul's hopeful vision for creation. In this passage, Peter[2] portrays a fiery judgment in which the heavens disappear, the elements are burned up and "everything" is "destroyed" (2 Pet 3:10-13). Why, to use Paul's language, would all of creation groan in longing for a future such as this?

One way to deal with the apparent conflict here would be to suggest that Peter and Paul simply disagree about God's promises for the future. A number of scholars adopt this apparently straightforward solution. But it is very unlikely that Peter would have thought that his portrayal of the future contradicted Paul's. Just a few verses later, in 2 Peter 3:15-16, he makes it clear that he holds "our dear brother" Paul's letters in the highest possible regard and considers them, in fact, to be among the Scriptures. Peter will surely consider the things that he has written about the promise of the Lord's return to be consistent with what Paul has already written in Romans and elsewhere. This fact alone challenges us to look for a different solution from one of irreconcilable conflict between Peter and Paul on this point. And for Christians who accept both Romans and 2 Peter as God's word for us, we have all the more reason to work to interpret Scripture in the light of Scripture. While we must be careful not to miss the unique message that each of these letters has for us, we are also right to seek coherence and consistency in what they reveal about Christian doctrine and to allow that which is clearer in one text to help us interpret that which is less clear in another.

The conviction of Scripture's essential unity in diversity is part of what drives us in this chapter to probe more deeply into just what 2 Peter 3 intends to tell us about creation's future. It prevents us from accepting without question the common notion that Peter's vision is of the annihilation of this creation and its replacement with something else, when Paul has made it clear in Romans 8 that this creation has a future in God's purposes. As it turns out, the popular caricature of the message of 2 Peter 3 proves inaccurate in any case when we take the time to explore beyond a superficial reading and consider the passage within its wider context. We will see that the apparent conflict between 2 Peter 3 and Romans 8 reflects primarily a difference in emphasis between these two texts rather than a substantive difference in eschatological vision.

BROKEN PROMISES?

We begin, as we always must, by looking at the wider context of 2 Peter 3:10-13. This will take us some way toward understanding why Peter's emphasis here is rather different from Paul's in Romans 8. In 2 Peter 3:3-4, Peter reminds his readers that he has previously warned them about "scoffers" who will come in the last days. The scoffers question whether God will ever visit his creation, whether there will ever be a final judgment. They say, "Where is this 'coming' he promised? Ever since our ancestors died, everything goes on as it has since the beginning of creation" (v. 4). These scoffers—who seem by the time of the writing of 2 Peter to be having a dangerous influence on the Christian community—doubt that God is ever involved in the world and claim that there will be no final reckoning when they might be judged for their actions. Hence they feel free to go on "following their own evil desires" (v. 3).

The scoffers sound a bit like the Epicureans of Peter's day who claimed that the gods exist in blissful separation from the transient world of humankind, such that there is no need for anyone to fear a future judgment; people should content themselves with the simple pursuit of pleasure in this life, seeking above all to avoid pain of any kind.[3] The scoffers also sound a lot like many people in our own day who, even if they believe in God, would rather that he did not actually involve himself in their lives and certainly would prefer that he never come to the earth in judgment. It is tempting for all of us to suppose that our actions do not really matter, to assume that there are no lasting consequences to how we live, especially when our lives and our actions are driven only by the desire to please ourselves. Our societies train us to become like the false teachers described earlier in 2 Peter: "experts in greed" (2 Pet 2:14), with an empty notion of "freedom" (2 Pet 2:19-20) that leads us to indulge ourselves at the expense of others, and ultimately at the expense of our own lives too.

Against the view of the scoffers, 2 Peter 3 sets the clear testimony of Scripture to God's acts in the past—in creation and in judgment—and points forward to the certainty of a future "day of the Lord" when all things will be laid bare before God. Just as the cosmos of Noah's day was deluged by the flood, so will the present world, according to Peter, face the fire of God's judgment (2 Pet 3:5-7). The apparent delay of the day of the Lord, which the scoffers claim invalidates God's promises, is not due to God's slowness, Peter says, but is due to his profound patience (2 Pet 3:8-9). God desires all to come to repentance, and his time is not like our time.

Given that Peter is addressing a context where scoffers deny the reality of God's judgment and claim that the world will go on forever as it always has, it is not surprising that Peter should find it necessary to emphasize the radical discontinuity that marks the day of the Lord. Peter makes it as clear as possible in this passage that things will *not* always go on as they have in the past; the world as we know it will be "laid bare" before the fire of God's judgment (2 Pet 3:10), and afterward there will be "a new heaven and a new earth" (v. 13). In the light of this, Peter emphasizes, we must radically reconsider what sort of lives we are to live (vv. 11, 14).

In Romans 8, Paul encouraged us to look forward to a new creation when our adoption as God's children is fully revealed, our groaning bodies are redeemed and the creation itself is freed from the bondage that it came under as a result of human evil. In 2 Peter 3, Peter challenges us to consider whether we are in fact living in the light of that coming day of the Lord, or whether our sinful actions actually contribute to the world's current plight and thus are deserving of the fire of God's judgment. Peter forces us to consider the significance of all that we do in the light of the radically different future that God has for his world; 2 Peter 3 reminds readers above all of the discontinuity that is necessary if the world is to become a place where righteousness is truly at home. It warns us against the complacency of assuming that everything will always go on as it does now.

THE DAY OF THE LORD

The language that Peter uses in his vivid description of what will happen on the "day of the Lord" (2 Pet 3:10-13) is borrowed from the Old Testament and from Jewish apocalypses that employ dramatic imagery to portray the salvation and judgment of God. This means that 2 Peter 3 is not the sort of text that we can read in the same way as we would a newspaper report, a historical narrative or a physics textbook. Like much of biblical prophecy 2 Peter 3 describes events that transcend ordinary human experience, and only metaphor, poetry and the language of apocalypse are adequate for the task. As we study this passage, then, we must keep in mind that though Peter intends to tell us about a real future event that represents the culmination of the world's history, we will miss the point if we try to derive from this text a "scientific" cosmology. Peter simply is not concerned with instructing us about the physical structure of the universe; his dramatic portrayal of the coming of God to his creation is meant instead to transform the way we live and act in the world today.

This passage is also not meant to give us a timeline for predicting the day of the Lord. In fact, Peter reminds us that any such attempt is futile. He cites the saying of Jesus recorded in Matthew and Luke that the Lord will come like a thief (2 Pet 3:10; cf. Mt 24:43; Lk 12:39), a motif that is also echoed by Paul (1 Thess 5:2, 4) and by John in Revelation (Rev 3:3; 16:15). In each occurrence it serves to highlight the unpredictable nature of the timing of Christ's return and the need for his followers always to be living in such a way that they are prepared for his coming. We will look in the next chapter at how Jesus uses this theme in the wider context of Luke 12. For now, however, our focus is on what Peter tells us about the day of the Lord and its consequences for all of creation.

According to 2 Peter 3:10, there are three things that will accompany the coming of the Lord to his creation:

"the heavens will disappear with a roar";

"the elements will be destroyed by fire";

"the earth and everything done in it will be laid bare."

Attentive readers who are familiar with older translations might know that the final event described here has sometimes been interpreted to involve the earth and its works being "burned up" (e.g., KJV) rather than, as in recent translations, "disclosed" (NRSV), "exposed" (ESV) or "laid bare" (NIV). The difference is due to the fact that different ancient manuscripts containing this verse have different Greek words here. Most recent commentators, however, are convinced that the reading "will be found" (Gk. *heurethēsetai*), which has by far the strongest manuscript support, is the original reading and so they would support something like the translation given above.[4] But this leaves us with the question: what does it *mean* for "the earth and everything done in it" to be "found" or "laid bare"? We will need to return to this question after considering the first two lines of Peter's description.

The first line refers to the disappearance of the sky above, accompanied by a "loud noise" (NRSV) or a "roar" (NIV). Peter describes this in verse 12 as the "destruction of the heavens by fire." Given that God is envisioned as enthroned in the transcendent "heaven" above (2 Pet 1:18; 1 Pet 3:22), this burning away of the earthly "heavens" suggests that the symbolic separation between God and his creation is being done away with; the earth is about to be visited by its Creator, Judge and Redeemer.

In the second line, Peter claims that the "elements" too will be destroyed with fire, because they will, according to 2 Peter 3:12, "melt in the heat." For readers today, this sounds like a description of the basic components of the material universe being destroyed. This is, in fact, one possible way that 2 Peter's first readers might have understood it too. The "elements" (Gk. *stoicheia*) would not for them be those of our periodic table, but they could refer to water, air, earth and fire as the principal elements of the cosmos. If this is the meaning of *stoicheia* in 2 Peter 3, this line would seem to refer to the burning up of the entire world, its final destruction in the fire of God's judgment. Verse 11, which begins, according to the NIV, "Since everything will be destroyed in this way," might seem to support such an interpretation. The NIV, however, misses a word in the Greek text at this point, and the NRSV translation "all *these* things" preserves the possibility that is present in the Greek of a more limited referent than "everything."

In any case, it is important to note that the "destruction" of the elements or "all (these) things" would *not* have meant for ancient readers the dissolution of the world into nonexistence.[5] The language of "destruction" is used in the Bible to describe something that has been rendered unfit for its purpose—a land made desolate by drought or flood, a burst vessel that no longer holds liquid, a ship's stern that waves have broken up, or even a coin that has been lost.[6] The idea is of something wrecked, ruined, broken apart or put beyond human use, not of something having been obliterated into nothingness. The parallel used by Peter himself in 2 Peter 3:6 helps us understand what he means: he compares the future judgment by fire with the past "destruction" of the world by the deluge of water at the time of Noah. The flood described in Genesis certainly could be said to have "destroyed" the world of the time. But once the waters receded, a dove was able to bring Noah a freshly plucked olive leaf from the drying earth (Gen 8:11). Later Jewish writers would consider the time after the flood to represent a "new creation" of sorts, but this was clearly a new creation that stood in material continuity with the world that existed before God's cleansing judgment. For 2 Peter, the final judgment of the world by fire represents an analogous—if more thorough and definitive—process of purification, renovation and renewal.

Before we go any further, however, we need to consider another possible interpretation of the elements mentioned in this passage. The Greek word translated "elements" (*stoicheia*) can sometimes be used to refer not to earthly elements (as we have assumed above) but to heavenly bodies, especially the stars, sun and moon.[7] These heavenly elements may in some cases even be associated with potentially evil spiritual forces.[8] As we will see, this meaning for *stoicheia* actually seems to fit the context of 2 Peter 3:10 rather better than the usual interpretation. A number of scholars have thus argued that Peter is describing here not the melting of the earth but the destruction of *heavenly* "elements" prior to God's judgment of the earth.[9] The idea expressed in 2 Peter 3:10 would in this case echo Isaiah 34:4, where an ancient Greek translation describes God's judgment as a time when "all the powers of heaven will melt."

Another ancient Christian writer in fact uses this Greek translation of Isaiah 34:4 to describe the fiery final judgment in a way similar to 2 Peter 3 and says that at the end of this judgment, "the secret and open works of humankind will appear" (2 Clement 16:3). Notice how, if we accept the meaning of "heavenly bodies" for *stoicheia*, 2 Peter 3:10 seems to describe the same sort of progression: (1) the outer heavens are torn away, (2) the intermediary heavenly bodies are dissolved with fire, and then (3) the earth itself and all the things done in it are laid bare before God, being "found" before him. There is no longer anything left to separate or hide human beings from the testing fire of God's judgment.

Peter is using vivid cosmic imagery in this passage to convey the common biblical idea that on the last day there will be nowhere for anyone or anything to hide from God's judgment. When the heavens are torn back and the "elements" destroyed by fire, the earth and all the works done in it will be disclosed before God's presence. Nothing will be hidden from God's sight as he acts to judge evil and to rescue his creation from its grip. Against the scoffers who think that they can avoid ever having their actions judged by God, Peter makes it clear that not only is God's judgment certain, but there will be no chance of covering up one's evil deeds on the last day. All will be brought into the light of God's truth and be disclosed before him.

The focus of God's judgment throughout this chapter is in any case clearly on *human* evil, not on the doing away with creation. For example, when Peter alludes in 2 Peter 3:7 to the fact that the present heavens and earth are "reserved for fire," he makes it clear that the fire is intended for "the day of judgment and destruction of the ungodly." This precise focus is probably why Peter stops short of describing the "burning up" of the earth itself. Once the heavens are figuratively torn back and the (heavenly) elements destroyed by fire in verse 10, the objects of God's judgment come clearly into view: the "works" that are done in the earth. Thus, when Peter summarizes the events of the day of the Lord again in verse 12, he mentions only that it will bring "the destruction of the heavens by fire, and the elements will melt in the heat." We have noticed that verse 11 seems to go further in calling this the destruction of "everything," but New Testament scholar Peter Davids points out that in the context this must refer once again only or primarily to those things just mentioned in verse 10: the "heavens" and the "elements." "Naturally such cosmic destruction would have an effect on the earth, as would the destruction of the ungodly," Davids says, "but that is not Peter's focus. Instead, his focus is the positive vision of the future and what it means for the present."[10]

Peter's "positive vision of the future" is "a new heaven and a new earth" (2 Pet 3:13). But Peter is keenly aware that it is only through the unmasking of human injustice and its judgment by God himself that this place "where righteousness dwells" can ever be realized. The dramatic scene of fiery "cosmic catastrophe" that he paints for us is intended to wake us up to the reality of God's judgment and to challenge us to consider the significance of how we live our lives now. His picture is similar to Paul's in 1 Corinthians 3:10-14, where Christian workers are warned to "build" carefully only on the "foundation" of Christ: Paul says that "their work will be shown for what it is, because the Day will bring it to light. It will be revealed with fire, and the fire will test the quality of each person's work" (1 Cor 3:13). In 1 Peter 1:7, we again find the same sort of challenge: here Peter encourages Christians to rejoice in their hope even in the midst of trials so that "the proven genuineness of your faith—of greater worth than gold, which perishes even though refined by fire—may [Gk. *heurethē*, "be found to"] result in praise, glory and honor when Jesus Christ is revealed." The fire of God's judgment is in all of these texts a purifying fire that destroys human evil, burning off the dross while renewing and purifying all that is built on the foundation of Christ and brings glory to God.

The "day of the Lord" described in 2 Peter 3 serves to judge human evil once and for all, and it challenges us to live radically different lives now, since all of the works done on earth will one day be revealed before God. It is striking that in Revelation 16:15 John describes the "great day of God the Almighty" (v. 14) in very similar terms: "Look, I come like a thief! Blessed is the one who stays awake and remains clothed, so as not to go naked and be shamefully exposed." In Revelation, to be unprepared for the day of the Lord is to be found (metaphorically) naked and shamefully exposed; to be ready for that day means being alert and clothed. Elsewhere in Revelation, we learn that such "clothing" is a pictorial way of referring to the "righteous acts of God's holy people" (Rev 19:8). God's people are expected to clothe themselves with righteousness in anticipation of the day when all will be revealed before him. Both Peter and John, then, like the Old Testament prophets before them and like Jesus himself, challenge us to pursue righteousness, to "live holy and godly lives" (2 Pet 3:11) in light of the future day of the Lord.

It is important to observe that this does not imply that we are made able to stand before God at the last judgment on the basis of our own merits. Revelation 19:8, for example, reminds us that even the righteous acts of God's holy people—the clothing of Christ's bride, the church—are "*given* her to wear." As we will see in chapter 8, the emphasis throughout Revelation is above all on Christ's atonement as the means by which God has redeemed his people. Peter too emphasizes that salvation comes only through repentance (2 Pet 3:9) and that we have "*received* a [precious] faith" only "through the righteousness of our God and Savior Jesus Christ" (2 Pet 1:1)—not by our righteousness, but by Christ's. Peter is, however, concerned throughout his letter that we not be "ineffective and unproductive" as we "confirm" our "calling and election" (2 Pet 1:8-11). We are to be on our guard and not fall from our position (2 Pet 3:17) as we live in expectation of the return of Christ.

How Then Shall We Live?

Just what sort of lives does Peter expect his readers to live in light of the coming day of the Lord? Negatively, our lives will not follow in the way of the scoffers, who indulge their own selfish desires in the belief that this world and its pleasures are all that matter. Nor will we buy into the false promise of "freedom" offered by those such as the false teachers whom Peter describes in chapter 2. These people "promise . . . freedom while they themselves are slaves of depravity" (2 Pet 2:19). Peter wants us to perceive the emptiness of such promises, to recognize that "people are slaves to whatever has mastered them" (v. 19). This so-called freedom, then, actually represents bondage to selfishness, to corruption and evil. Genuine freedom is found only in entrusting ourselves to our Savior Jesus Christ, who enables us to resist the corruption of the world and to follow in "the way of righteousness" (2 Pet 2:21).

Our own culture presents us with plenty of its own unbiblical promises of "freedom"—freedom to do whatever I like, however I like, whenever I like, wherever I like, whatever the consequences for my neighbors and for the rest of creation. We and most readers of this book are no doubt deeply grateful to live in societies and under governments that grant us substantial religious and political freedoms. But 2 Peter challenges us to resist the temptation of buying into our culture's idolatrous exaltation of individual liberty above all else—above the claims of other human individuals and the claims of nonhuman creation, the less powerful voices of all those whom our culture would urge us not to hear. In how many ways are we inclined to ignore the consequences of our decisions for others when we are promised ease, convenience, wealth and freedom! In the face of such temptation, Peter points us instead to Christ, to the one who gave up his very life on behalf of others, for creation itself—for us. How can we claim to be followers of this Lord and Master and not follow in the same path?

The way we resist the allure of such false promises is by following Jesus, living "holy and godly lives" as we "look forward to the day of God and speed its coming" (2 Pet 3:11-12). Do not miss how extraordinary this statement is. Peter has just reminded readers of the absolute unpredictability of the day of the Lord and the fact that its origin is not from us but from God himself. Yet here Peter suggests that we might actually be able to "speed" the coming of the day of the Lord. How is this possible?

It would seem that Peter, like the rest of the New Testament writers, envisions God's grace as encompassing even the works that he gives us to do. God includes us in his purposes and makes us able to participate in the work of his kingdom. We do not bring about the kingdom of God: it is wholly God's gift. Yet God predestines and enables those who are in Christ to live now as citizens of that kingdom. The promise of resurrection and new creation means that our earthly labor is never in vain in the Lord (1 Cor 15:58), however much it may sometimes seem that way to us. We live in expectation of God's promise to bring in the new creation, living now as those who are Christ's subjects, renewed through the Holy Spirit and enabled to do work that has genuine value before God. In this passage, Peter merely expands the New Testament picture even further by suggesting that God actually uses our efforts now to speed the day of his coming.

The "holy and godly" way of life (2 Pet 3:11) that God uses to accomplish this is not one of pious remove from the dirty, difficult and dangerous challenge of living in this broken world. It is not, to be sure, a way of life that is invested in the world as it now is, oriented toward the expectation of reaping worldly returns from things that are passing away. But it is nonetheless a way of life that has profound consequences for the world as it now is, because it is a way of life that embodies God's priorities and is lived in the expectation of "a new heaven and a new earth, where righteousness dwells" (2 Pet 3:13).

VIRTUES FOR A NEW CREATION

The Greek words that lie behind "holy and godly lives" in 2 Peter 3:11 are plural and might better be translated "holy ways of living and godly acts." Such a way of life comprises all sorts of specific behaviors and individual acts. In chapter 1, Peter has in fact listed some of the attributes that mark such a life:

> For this very reason, make every effort to add to your faith goodness; and to goodness, knowledge; and to knowledge, self-control; and to self-control, perseverance; and to perseverance, godliness; and to godliness, mutual affection; and to mutual affection, love. For if you possess these qualities in increasing measure, they will keep you from being ineffective and unproductive in your knowledge of our Lord Jesus Christ. (2 Pet 1:5-8)

To be effective and display true knowledge of Christ, Peter tells us that we must "make every effort" to develop these virtues. Peter has just reminded readers that we have in fact already been given "everything we need for a godly life" (2 Pet 1:3), so God's gift and our work are not here mutually exclusive; his grace includes our pursuit of these virtues. It will do no good, then, for us to complain that these attributes are beyond our abilities or too difficult for us; it is God who enables us to embody them in our lives and actions.

If it is God's will to do this work in us, we also need not respond with the sort of anxiety and fear that led Felix, for example, to dismiss Paul and refuse to listen any longer when Paul spoke to him about "righteousness, self-control and the judgment to come" (Acts 24:25). As we will see in the next chapter, there is a place for proper fear of the Lord: "The fear of the

LORD is the beginning of wisdom" according to Proverbs 9:10, and Moses tells the Israelites, "And now, Israel, what does the LORD your God ask of you but to fear the LORD your God, to walk in obedience to him, to love him, to serve the LORD your God with all your heart and with all your soul" (Deut 10:12). Paul too encourages the Philippians to "work out your salvation with fear and trembling, for it is God who works in you to will and to act in order to fulfill his good purpose" (Phil 2:12-13). We have seen that one of the purposes of the letter of 2 Peter is to wake readers up to the reality of God's judgment. But there is a difference between a proper fear and reverence of the Creator, Lord and Judge of the universe and the sort of fear that leads us to turn away from God, or to forget that his judgment and his wrath in the face of evil and injustice is an expression of his love, or to ignore the salvation that he provides in Christ. We are not, according to 1 Peter 3:14, to fear what others fear, because, as Proverbs 19:23 puts it, "The fear of the LORD leads to life; then one rests content, untouched by trouble."

Fear of the Lord, then, is just the opposite of the sort of self-focused anxiety that can prevent us from heeding the challenge of a text like 2 Peter. Rather, like the early church, "living in the fear of the Lord and encouraged by the Holy Spirit" (Acts 9:31), we can entrust ourselves to God and his grace, finding that "perfect love drives out fear" (1 Jn 4:18). Self-control then becomes a real possibility for us because we are freed from the power of sin and death and enabled to see clearly the deceitfulness of our society's hollow promises of freedom. The anxiety that holds us back from opening our hands and hearts in genuine love for others is overcome by the peace of Christ.

In a traditional list of virtues such as in 2 Peter 1, incidentally, Peter does not mean that we ought to add these virtues one by one, as if we wait to pursue love until we have perseverance figured out. Rather, they are mutually supporting and intertwined, all integral parts of what a Christlike life looks like. In a world facing the sorts of profound challenges surveyed earlier in this book—ignored or downplayed by many, but causing despair for those most aware of them and most involved in trying to address them—we surely are called to reflect deeply and at length on what it will look like for us to embody such virtues today. How do people of *goodness* or *virtue* (*arête*) and *godliness* deal with the growing knowledge that our collective actions are destroying so much of the diverse life created by God and at the same time potentially condemning millions of our fellow human beings to poverty, starvation and death? What will it mean for us to exercise genuine *self-control*, not only in our sexual lives (so often the exclusive focus of our teaching) but also in our economic lives, in how we do business, in what we consume and in what we choose not to consume? What will we decide is "enough" if we truly accept that God has given us all we need and that he desires us to reflect his generosity in how we live among our fellow creatures? What will it look like to *persevere* if, despite all our efforts, nothing seems to change, or if today's secular prophets of doom turn out to be right and things actually get much, much worse?

We will not always have precise answers for these questions, and some answers will be necessarily tentative, reflecting not only our always incomplete knowledge of God and his purposes but also our imperfect knowledge of the world and the challenges it faces. The Bible is not, after all, a rule book, and we require wisdom to live and act rightly. We must acknowledge too that Christians are called, in the many different contexts where God places us, to a variety of ways of living out the "holy and godly lives" demanded of us. And for all of us the pursuit of these virtues will be an ongoing process of seeking to embody them "in increasing measure" (2 Pet 1:8). We remain always dependent on God's grace and mercy. But we must not let such caveats blunt the radical challenge of 2 Peter. If we miss the power of these texts to destabilize our comfortable lives and to reshape us into Christlike people who are prepared to follow in the way of the cross, to enter into self-sacrificial work on behalf of our neighbors around the world and of an entire groaning creation, we have truly become "ineffective and unproductive" in our knowledge of the Lord Jesus Christ (2 Pet 1:8).

A Place Where Righteousness Makes Its Home

Peter culminates his description of the day of the Lord with a focus on the hope of new creation that it brings:

> But in keeping with his promise we are looking forward to a new heaven and a new earth, where righteousness dwells.
>
> So then, dear friends, since you are looking forward to this, make every effort to be found spotless, blameless and at peace with him. (2 Pet 3:13-14)

We will in chapter 8 consider in more detail the biblical promise of "a new heaven and a new earth," which echoes the Genesis creation narrative, is developed in Isaiah 65–66 and is expanded most fully and beautifully in Revelation 21–22. It reminds us first and foremost that the Christian's ultimate hope is not for a disembodied existence apart from creation but for new embodied life in a new creation, a creation with a heaven *and* earth. Peter does not develop the picture at length for us here (perhaps assuming that his readers already know something about the biblical promises of a new heaven and new earth), but what he does choose to highlight is striking.

The new creation is a place "where righteousness dwells." The image echoes Isaiah 32, where, after God judges human evil in the land, "the Spirit is poured on us from on high" and "the desert becomes a fertile field, and the fertile field seems like a forest. The LORD's justice will dwell in the desert, his righteousness live in the fertile field" (Is 32:15-16). "The fruit of that righteousness," Isaiah 32:17 goes on to say, "will be peace"—*shalom*. Notice that 2 Peter 3:14 similarly encourages readers, as they await the new heaven and new earth, to be found "at peace with him." The hope for the future held out by 2 Peter involves, as we have seen, a radical break with the past. But in the end Peter's vision evokes the rich ecological vision of Isaiah 32, where the wilderness, cleared of human injustice, is renewed by God's Spirit as a fruitful field where righteousness dwells and peace is found. The challenge to 2 Peter's readers is to live now, in the present, as members of that community of peace, as those who have entered already into God's *shalom* even as we await the coming of our Lord and the fullness of the new creation that will accompany his coming.

Peter's Enduring Challenge

In our present context of environmental crises, whether real or imagined, present or predicted, Peter reminds us that our world's fundamental problem is not technological but moral. Human sinfulness, evil and injustice must be addressed if the earth is ever to become the place God intends it to be and we long for it to be. God's purposes for his creation and for his people will not finally be realized unless and until all things are put to rights. This putting to rights requires a radical rupture with the past, God's definitive judgment of evil, and his creation of a new heaven and a new earth. We observed in chapter 4 that many of us today are uncomfortable with a God who acts in the world in judgment, and we are often unwilling to take seriously our own sinfulness and need for grace and mercy. The claim of 2 Peter, however, is that this is where we must begin (and end) if we are to learn to live in a way that is in keeping with "a new heaven and a new earth, where righteousness dwells."[11]

7

Jesus, A Thief in the Night and the Kingdom of God

> *Till Armageddon, no Shalam, no Shalom.*
> *Then the father hen will call his chickens home.*
> *The wise men will bow down before the throne,*
> *And at his feet they'll cast their golden crown,*
> *When the man comes around.*
>
> JOHNNY CASH, "WHEN THE MAN COMES AROUND"

◆

PREPARING FOR THE UNPREDICTABLE

When Peter compared the Lord's return to the coming of a thief in the night (2 Pet 3:10), he was not inventing the analogy. As recorded by both Matthew (24:43) and Luke (12:39), it was Jesus who first used this image to describe the unpredictable nature of the coming of the Son of Man. In a book focused on Christian hope, a hope that is centered on the person of Jesus, it is worth taking a chapter to consider the significance of what Jesus himself said about hope and the future. It is not that the testimony of the other biblical texts we have examined are any less God's word than the Gospel writers' record of Jesus' words during his earthly ministry. But the challenges and assurances that accompany Jesus' description of the return of the Son of Man and his promise of the kingdom of God are profoundly relevant to the theme of this book, to what it might look like in our present situation to live in light of the future promised us in Christ. Luke 12 is particularly helpful in the way that it brings together several strands of Jesus' teaching on life in the kingdom, and so this chapter will serve as a focus for our brief study. Beginning with Jesus' comparison of the return of the Son of Man to a thief in Luke 12:39, we will work outwards from this verse to situate Jesus' warning within the context of the rest of the chapter. This approach has the added benefit of providing us with the opportunity to consider whether the faithful household "manager" or "steward" whom Jesus describes in Luke 12:42-46 provides us with any guidance for our "stewardship" of the earth.

In Luke 12:39-40, Jesus makes the following observation: "If the owner of the house had known at what hour the thief was coming, he would not have let his house be broken into. You also must be ready, because the Son of Man will come at an hour when you do not expect him." Jesus here makes two things clear: (1) it is impossible to predict when the Son of Man will come, and (2) his followers must always live in a state of readiness for his imminent arrival. Who is this "Son of Man"? It is evident from other passages that the "Son of Man" refers to Jesus himself (e.g., Lk 5:24; 7:34; 9:22, 44; 18:31; 22:22, 48; 24:7), but the precise significance of this self-designation has been the subject of much scholarly debate.[1] Jesus probably used the expression in part because of its potential ambiguity: it challenged his listeners then (and all later readers of the Gospels) to consider the significance of just who Jesus is. On the one hand, the phrase can be a general way of referring to "someone" or even to humanity in general; in this sense, it calls attention to Jesus' shared or even representative humanity. But when, as here, Jesus talks about the "coming" of the Son of Man, and especially his coming "in glory" (Lk 9:26) or "in a cloud" (Lk 21:27), he is making a more radical claim, associating his identity as the "Son of Man" with the exalted, heavenly figure of Daniel 7. This mysterious figure, "one like a son of man," comes on the clouds of heaven before the Ancient of Days and is given "authority, glory and sovereign power," and all peoples of the earth worship him (Dan 7:13-14). His rule is said to be everlasting, and his kingdom is "one that will never be destroyed" (Dan 7:14). When Jesus talks about the coming of the Son of Man, then, he is also talking about his lordship over all the earth and the dramatic arrival of his universal kingdom.

Based on what Jesus claimed about the future coming of the Son of Man (as recorded in Luke 12 and elsewhere), his followers looked forward to his eventual return in glory after his death, resurrection and ascension. Immediately after his ascension Jesus' disciples were told, "This same Jesus, who has been taken from you into heaven, will come back in the same way you have seen him go into heaven" (Acts 1:11). No doubt many of Jesus' earliest followers expected him to return within their lifetime. Yet, as Jesus emphasizes in Luke 12, no one could know the timing of his coming—not even Jesus himself (Mt 24:36; Mk 13:32). This certainly means that we can safely ignore anyone who claims they can predict the time of Christ's return, but more importantly it challenges all of Jesus' followers to alertness and readiness, to a life lived in eager expectation of the imminent return of our Lord and King. This, in fact, is the main point of Jesus' parables in Luke 12.

WAITING FOR THE MASTER

It is risky and potentially misleading to attempt to squeeze significance out of every detail of Jesus' parables (as if they were all extended allegories), because the details are often incidental to his main point. The "Son of Man," for example, is like a "thief" in the unexpectedness of his coming, but not in his character or his actions! Nonetheless, it is certainly not incidental that the two parables about readiness that Jesus tells in Luke 12 (cf. Mt 24:42-51; Mk 13:33-37) are about servants awaiting the return of their master (*kurios*, "lord"). This, after all, is the situation in which God's people find themselves as they wait for the coming of their Owner and Master, the Lord (*kurios*) of all the earth.[2] God is not, to be sure, absent from his creation in the way that the master of these stories is away from his household; he remains actively at work in the world and in us. Yet the point of the parables that Jesus tells in Luke 12 is to emphasize the way in which our expectation of the future triumphant return of Christ ought to transform how we live in the present. "It will be good," Jesus says, using the same word, *makarios* ("blessed"), that he uses in the Beatitudes, "for those servants whose master finds them watching when he comes" (Lk 12:37). "It will be good," Jesus emphasizes again in the next verse, "for those servants whose master finds them ready," even if he comes at an unexpected hour (Lk 12:38). Jesus says his disciples must always "be ready," because the Son of Man will come like a thief in the night, at a time when he is not expected (Lk 12:40).

How are Jesus' followers meant to get ready for the return of the Son of Man? Such preparation involves living as members of God's kingdom now, as those whose Master and King is God in Christ, even when that kingdom seems far off and when it seems like other powers are in control. The Son of Man will come as Lord over all the earth, as the one who has been given dominion and an everlasting kingdom. To live in lively expectation of his coming—and with the conviction that he is the true Lord and Master of all—means living now as those whose lives are marked by the priorities of his kingdom.

Just as we discovered of Paul's portrayal of the "new creation" and our identity as God's children in Christ, Jesus' announcement of the kingdom of God involves an "already and not yet" dynamic. Thus, when the Pharisees ask Jesus when the kingdom of God will come, Jesus first tells them that "the kingdom of God is in your midst," but they have been unable to observe it (Lk 17:20-21). Yet Jesus then immediately goes on to tell his disciples about the unmistakable signs of the *future* coming of the Son of Man: "For the Son of Man in his day will be like the lightning, which flashes and lights up the sky from one end to the other" (Lk 17:24). The kingdom is here already in the person of Christ the King, and his power and victory over sin, death and evil are displayed above all in his death, resurrection and ascension; but his servants yet await his return in glory to judge and to give new life to all he has redeemed. In this in-between time, Jesus' followers are called to watchfulness and readiness.

Jesus' words and deeds throughout the Gospels reveal to us what such watchfulness looks like—what the kingdom of God is like, and what it means to live as his subjects now, as those who eagerly await his return. In the immediate context of Luke 12, what we learn about life in the kingdom is especially that it involves responsible, selfless stewardship of all that has been entrusted to us. It is in this chapter that we encounter the image of the disciple serving as a steward or manager of what belongs to the master, both of his other "servants" (Lk 12:42) and of all his "possessions" (Lk 12:44). Jesus uses this image of a steward in a reply to Peter's question about whether the challenge to be ready for the unexpected coming of the Son of Man applies to everyone or just to the disciples (Lk 12:41). Jesus' response suggests that this teaching does indeed apply to everyone, but that there is an even greater responsibility for those who have been given special privileges, for those who have been granted intimate knowledge of the master's will. Jesus concludes with the stern challenge: "From everyone who has been given much, much will be demanded; and from the one who has been entrusted with much, much more will be asked" (Lk 12:48).

This challenge is one that seems directed especially at Jesus' disciples, as those who have been entrusted with intimate knowledge of who Jesus is and what is expected of them. They bear particular responsibility for the people over whom they will be leaders and, as James says of teachers generally, they "will be judged more strictly" (Jas 3:1). Yet this challenge is also one that ought to resonate especially with all of us who are followers of Christ today in the richer nations of the world. We have, first of all, along with all believers, been entrusted with God's Word, which supremely reveals his purposes for us. We live, moreover, in an age when the resources for studying Scripture are more abundant and more readily available than they ever have been. We have also been given immense benefits—tangible and intangible—that derive from the freedom, wealth and power of the societies and countries in which most readers of this book happen to live. We also live in an age when not only are the effects of our collective actions on other people and the rest of creation dramatically greater than they ever have been, but—and we must count this as a gift, however awkward it makes life for us now—we are uniquely able through science and technology to quantify and understand much about the nature of these effects.

We have no excuse for remaining ignorant of the consequences of our actions. It is not surprising that we often experience such knowledge as a wearisome burden: so many everyday things that in former times appeared morally neutral—what we eat, what sort of home we live in, how we keep warm, where and how we dispose of our garbage, how we get around—are now fraught with moral weight. Yet the very fact that most of us can make choices about such things attests to our relative freedom and wealth. We cannot help but hear Jesus' challenge to his disciples as one that applies to us: from those who have been entrusted with much, much is demanded.

LORD OF THE HARVEST?

Jesus has provided a negative example earlier in Luke 12 of what it looks like to reject this responsibility, to squander what is given to us and fail to be "rich toward God" (Lk 12:21). He tells a parable about a rich man, an abundant harvest and what the rich man decides to do with the harvest. The parable begins, unusually, not with the human character as the subject but with the land itself.[3] "The ground of a certain rich man," Jesus says, "yielded an abundant harvest" (Lk 12:16). Whatever work the farmer (or his servants) may have done to plant, nurture and harvest the crops, it is ultimately the land that yields its produce, the soil that provides the nutrients, the rain that provides the water, the sun that provides the energy. Behind all of this is the beneficent Creator, the true owner and sustainer of the land. Yet the focus of the rich man is entirely on himself; even his conversation is directed only to himself! His assumption is that these are "my crops," "my surplus grain," all of which belongs to him and which will enable him to "take life easy; eat, drink and be merry" (Lk 12:17-19).

Of course, things do not turn out that way. Just as the Son of Man returns at an unpredictable time, so too, Jesus reminds us, does the time of our death remain outside our control and beyond our ability to predict. God tells the man that he is a "fool" and that his life is demanded of him that very night. The rich man's plans for himself have come to naught, and he is left with no control over the things that he had thought to prepare for himself. "This is how it will be," Jesus says, "with whoever stores up things for themselves but is not rich toward God" (Lk 12:21).

Jesus' parable leaves open the question of what the man ought to have done with his surplus harvest, of just how he could have been rich toward God. Jesus will provide something of an answer in the next passage. But here, above all, it is the self-absorption, the selfishness and the greed of the rich man that stand out as deserving of critique. His plans are wholly for himself and his view of possessions is that they are his to dispose of how he likes. He acts, therefore, as a "practical atheist," ignorant of his Creator and Lord. James describes similar sorts of people, those who say, with no thought of what might be the Lord's will for them, "Today or tomorrow we will go to this or that city, spend a year there, carry on business and make money" (Jas 4:13). They forget, James says, that their life is "a mist that appears for a little while and then vanishes" (Jas 4:14). Life lived for oneself, cut off from the Creator, Redeemer and giver of life, the true Lord and owner of all things, is an insubstantial thing, here today and gone tomorrow.

What prompts Jesus to tell the parable of the rich fool is a request from someone in the crowd that Jesus arbitrate between him and his brother in a dispute about an inheritance. Clearly recognizing Jesus as someone with authority, the man from the crowd says to him, "Teacher, tell my brother to divide the inheritance with me" (Lk 12:13). It sounds like a plea for justice, and we might expect Jesus to sort things out, to make sure that the right thing gets done. But if that is what we expect, Jesus surprises us. He is less concerned in this situation with what the man in the crowd may or may not deserve to receive from the inheritance than he is with the man's focus on possessions in the first place. Rather than acting as judge between him and his brother, Jesus warns him and all of those who are present (the imperatives are plural), "Watch out! Be on your guard against all kinds of greed; life does not consist in an abundance of possessions" (Lk 12:15).

It is almost too obvious to observe the way in which Jesus' warning strikes at the heart of our own consumerist cultures, yet we probably cannot be reminded too often of the counterclutural priorities of God's kingdom. With our eyes so often distracted by the false but enticing visions of the "good life" paraded by entertainment, media, advertising, friends and even sometimes our churches, and with a deep-seated impulse to claim that we deserve whatever it is we want, we do well to return again and again to the radically different vision of life in God's kingdom that Jesus gives us. If our life is not, in fact, defined by our possessions, and all that we have belongs truly not to us but to God, what difference does it make to our understanding of justice, to our understanding of what we "deserve" because we have worked so hard? What difference does that make to our goals and aspirations for ourselves and for our children? What difference does it make to how we understand the so-called sacrifices that are necessary if we are to provide space for the flourishing of other people and all of creation? What difference does it make to how we understand human flourishing in the first place?

These are not new questions, nor are they unique to considerations of how Christians today care for creation. Yet the medium-term future of earth, to the extent that God has entrusted it to us, has much to do with how we answer such questions—and how we actually live in light of our answers. Historians have sometimes blamed Christianity for our environmental crises, claiming that the Christian view of "dominion" means that we have treated creation as something that exists only to serve our own needs and desires. It is worth speculating whether there would be any basis for such a critique (however unfair it may be in any case) if the Christian church had always managed to live faithfully and consistently the sort of life that Christ tells us God's kingdom consists in—even *apart* from any explicit recognition of our responsibility to care for God's creation.[4] I (Jonathan) once gave a talk in a Christian university about our responsibility to care for creation, where, despite my best efforts, an audience member remained rather skeptical about including creation care alongside the church's priorities of gospel proclamation and care for the poor. Yet, as I was chastened to discover later, this skeptic embodied in his own Christ-centered, generous and humble lifestyle a way of living that was far better than mine in its effects on the earth and other people.

Now the reality is that many of the global challenges described in chapters 2 and 3 of this book cannot be adequately addressed without some major economic, technological and societal changes. So just as some Christians spoke boldly in the past to end the slave trade and some work tirelessly today in the public arena to advocate for such things as debt forgiveness, poverty alleviation, care for orphans and respect for the lives of unborn babies, so must many Christians today be involved in local, national and global efforts toward a more just and sustainable world. We all ought to find ways to support those who are actively caring for creation for

God's glory. Yet change must also begin with each of us individually, here and now, in our own lives; and for this we have much to learn from traditional Christian piety, wisdom and virtue, rooted as it is in the radically countercultural vision of God's kingdom inaugurated by Christ.

FEAR AND TREMBLING?

For a book about hope, this chapter and the last have had rather a lot to do with fear: fear of the return of the Master, fear of the consequences of the coming of the Son of Man, fear of judgment. As we observed in the previous chapter, this is because there is undoubtedly some sense in which fear plays a role in the motivation expected of Jesus' followers. It is all very well for us to argue that fear is not an "ethical" way to motivate good behavior, but Jesus claims that there is indeed a sort of fear that is right and proper for us. "Fear him who, after your body has been killed, has authority to throw you into hell," Jesus says in Luke 12:5 (and see again Ps 111:10; Prov 1:7; 9:10). There is a proper fear and reverence of the one who is Creator, Owner and Judge of all the earth, of the one to whom we owe our allegiance and our lives.

Yet even as Jesus encourages a right and proper fear of God in Luke 12, he reminds us immediately of the character of this God whom we are to fear. He is the one who cares for every single sparrow—"not one of them is forgotten by God" (Lk 12:6). "Indeed," Jesus says, "the very hairs of your head are all numbered. Don't be afraid; you are worth more than many sparrows" (Lk 12:7). Fear and reverence of such a God thus means *not* being afraid of anything else; it means *not* being afraid of the one who is merely able to kill your body (Lk 12:4); it means *not* being fearful and anxious for your own life. Jesus reveals to us a God who cares for even those small and common creatures that are rarely valued highly by human beings—and a God who cares all the more for us.

NO WORRIES

This becomes the great theme of Luke 12:22-34 (cf. Mt 6:25-34), which in many ways stands at the center of this chapter as the positive counterpoint to the warnings of Luke 12:35-48 and the negative example of the rich fool in Luke 12:13-21. If the story of the rich man showed the foolishness of storing up wealth for oneself and failing to be rich toward God, Jesus addresses in Luke 12:22-34 what is often at the root of such greed and avarice: fear and anxiety. It is so often fear for our well-being, fear for our children, anxiety about having enough and anxiety over what tomorrow may bring that prevent us from the sort of open-hearted and open-handed generosity that Jesus says marks life in God's kingdom. Our own culture's fears about the future and growing despair about the prospects for life on earth may well tempt many of us to turn inward, to hold on to what we have and to refuse to give up anything for the sake of others. In the place of worry and anxiety, however, Jesus encourages a steady trust in the Creator and Sustainer and an active seeking after his kingdom. In the place of acquisitiveness and storing things up for ourselves, Jesus tells his disciples to sell their possessions and give to the poor (Lk 12:33). "Seek his kingdom," and God will provide all we need as well (Lk 12:31).

Jesus turns to the natural world to help his disciples understand what it means to live like this, to live in ongoing dependence on the Creator. The rich fool of the parable forgot his dependence, forgot that the land and its produce was a gift of the Creator, forgot that he was not the true owner of any of it, and so deluded himself into thinking he could store it all up for himself. The birds, by contrast, do not sow or reap (Lk 12:24). Unable to store food, they live in ongoing, day-by-day dependence on God's sustaining provision through his creation. Jesus' disciples, who are of greater worth than birds (and notice again the assumption both that nonhuman creatures have value before God and that human beings are of greater value), ought not to be anxious and tempted to store up things for themselves. Rather, they must learn to trust in the beneficence of the Creator. Jesus directs his disciples' attention to the wildflowers of the field, which, though of apparently only passing significance, here today and gone tomorrow, are clothed in greater splendor than even the famed Solomon. For the disciples too, Jesus says, God will provide all that they need. Jesus is not, it should be clear, arguing that his followers should stop working for their food and clothing (after all, birds do "work" for their food). Nor is he issuing a policy statement about crop production and storage. Jesus is, however, radically challenging his disciples to entrust themselves to the care of God and therein to find the spacious freedom that enables them to participate in and enjoy the generous life of the kingdom.

It is sometimes queried how Jesus' followers in times of famine and want could possibly read this parable without irony or bitterness. In the light of a gracious God who cares and provides for his creation, how do we make sense of a world so often beset by hunger—or indeed of a world full of beleaguered, suffering and even dying creatures and ecosystems? Though we will never in this life find entirely adequate answers to such questions, Jesus actually provides us with one response right here in this passage. Jesus himself assumes that many in this world do *not* in fact have enough, that there are those who are poor and in need of help. Yet trust in God's provision ought to free those who do have more than enough to give freely and generously to those who do not. "Sell your possessions and give to the poor," Jesus says (Lk 12:33). This is how to "store up" real, lasting treasure, a treasure kept safe in heaven. This is what it is to be rich toward God. This is how to ensure that our hearts do not become hopelessly entangled with our possessions and with our schemes to keep things for ourselves, but rather that the affections of our hearts remain focused on the glory of God in Christ. This is how we seek God's kingdom.

The best commentary on Luke 12:33, though it is not necessarily directly dependent on it, is found in the *Didache*, or "Teaching of the Twelve Apostles," one of the very earliest Christian documents that we have apart from the New Testament. Readers of this early summary of Christian life and practice are instructed to give freely to those who ask, without any expectation of repayment. What is striking is the reason why they are to give: they are to give because "the Father desires to give to all from his own gifts" (*Didache* 1:5, our translation). God wants to give generously to all, and the means he chooses to accomplish this is through the generous giving of his people, whose possessions properly do not belong to themselves but remain always God's "own gifts." God has entrusted us with the task of bringing about his will that the abundance of his creation be shared with all; and as with everything that he calls us to do, it depends first and last on his grace, on his own generous provision for us.

Life in the Kingdom: Love, Stewardship and Hope

This brings us back again to what is central to life in the kingdom of God: love. The "royal law" of love (Jas 2:8), or "law of the kingdom," is famously encapsulated by Jesus in his combination of Deuteronomy 6:5 and Leviticus 19:18, cited in Luke 10:27 (and Mt 22:37-39; Mk 12:30-31): "'Love the Lord your God with all your heart and with all your soul and with all your strength and with all your mind'; and, 'Love your neighbor as yourself.'" Jesus' teaching in Luke 12 has shown us how such love for God and neighbor radically challenges our usual understanding of possessions and how we should prepare for the future. And though we have not drawn out all the implications of this for our care of creation, it is now abundantly clear that we cannot care for our global neighbors—or even truly for our neighbors next door—if we do not also care for the rest of creation of which we are all a part. It must also be evident, as we observed in chapter 4, that if we love God we will value and care for all that he values. The ethics of the kingdom that we proclaim, then, extend beyond even our care for the poor and for all our human neighbors to embrace the whole of creation under the lordship of Christ.

Many Christians have found the image of a steward, like the "faithful and wise manager" (or "steward," Gk. *oikonomos*) cited by Jesus in Luke 12:42, to be useful for describing the nature of these responsibilities that we have within creation. "Stewardship" has in fact become the predominant way in which our relationship to creation is described, even in many cases by non-Christians, who speak of stewarding creation for future generations. There are some who argue, however, that the term has lost any usefulness it may once have had in the process of becoming so widely accepted. Its very ubiquity means that faithful and wise stewardship is interpreted in almost any way someone likes, including in ways that justify what most of us would characterize as nothing other than self-serving exploitation of the earth's resources. Others argue that it is not just abuse of the term that is the problem, but the notion of stewardship itself is inherently flawed in its connotation of an exclusively top-down, managerial approach to conserving, allocating and using "natural resources."[5] This version of stewardship treats God's creation as nothing other than a larder full of resources to be used exclusively for the benefit of ourselves and our offspring, rather than as a living, dynamic network of interconnected life that brings praise and glory to God and of which we are an integral part.

Nonetheless, stewardship can remain a useful way of conceiving of our responsibilities within creation if we keep the following principles in mind, all of which helpfully echo elements of Jesus' parable of the faithful and wise manager in Luke 12:42-48: (1) our place within the household of creation is distinct from that of other creatures, but we are nonetheless a part of the same household, fellow servants of the same Master and entrusted with unique responsibility for the well-being of the rest of his household; (2) God, not us, is both the Master and the Owner of the creation that we steward, and he continues to be intimately involved in his creation; and (3) our stewardship of creation is therefore done first and foremost not for our own sake, or for the sake of future generations, but above all for the sake of God and his glory. In this way, our stewardship of creation becomes an expression of our love of God, of our faith in Christ and of our hope for his kingdom.

Such a description of stewardship is really just a way of clarifying how we best exercise the dominion given to us by God in Genesis 1. Our God-ordained "rule" over other creatures and "subduing" of the earth (Gen 1:26-28) has of course sometimes been interpreted in purely exploitative terms. Yet such an interpretation runs counter to the way kingship is conceived of in the Old Testament, where the ideal king rules as the first among equals and for the benefit of those ruled. To rule over all the other creatures in this way is God's intention for humankind, reflecting our proper place within creation as the bearers of his image. As we will see in the next chapter, this royal role is sustained and renewed even into the new creation, where those who are redeemed by Christ are said to "reign on the earth" (Rev 5:10; 22:5). The even stronger language of "subduing" the earth used in Genesis 1:28 tells us that this rule includes such things as the active working of the ground. Anyone who has done any farming (or even just sought to grow their own vegetables) will recognize the appropriateness of the language of "subduing" for describing what is involved in working with the earth to produce food. In the next chapter of Genesis we learn in fact that even in Eden there was such work to be done. The Lord placed the man in the garden "to work it and take care of it" (Gen 2:15). The Hebrew word for "work" here is used in other contexts for the work of the priests in God's temple, and so we have a hint here that our working of the earth is intended to be a part of our service toward God, its true Owner and Sustainer.

Alongside such work in any case is the command to "care" or "keep" or "protect" the earth, as we might variously translate the text here. This command reminds us of the emphasis throughout Genesis 1 on the goodness of all creation; it is valued by God and so must be valued—and cared for—by all those who love him. The rule and the work of God's image bearers ought always, then, to tend toward the health and preservation of his good earth. If Christians were in any doubt about what such rule looks like, we find in Christ, the true image of God and the one through whom God's image is renewed in us (Col 1:15; 3:10), a picture of what rule on behalf of others looks like. Indeed, when we consider the example of Christ and his sacrificial love, it is understandable why some Christians find even the notion of stewardship insufficiently radical for describing the nature of our responsibilities within and for God's creation.

We must end this chapter, however, with yet another reminder that this profound responsibility to steward or care for creation does not mean that we are left alone in our task or that in the end it is all up to us. God is not an absentee landlord. God the Father remains at work in his creation (Jn 5:17). The Holy Spirit, as we observed in Romans 8, helps us in our weakness, gives us life and—as Jesus encourages us in Luke 12:11-12—even teaches his followers what to say when they are brought before synagogues, rulers and authorities. And at the end of Matthew's Gospel Jesus assures his disciples, just after giving them the Great Commission, of his ongoing presence with them: "And surely I am with you always, to the very end of the age" (Mt 28:20).

God remains at work in and through us even now to bring about his purposes in his creation; and we are promised too that the kingdom for which we work is finally and entirely his gift. Thus we return to a quote from Luke 12 that we included at the beginning of chapter 4, where we set it alongside the popular but unbiblical claim that the future is achieved only by our own efforts. Jesus, we have seen, hardly lets us ignore our profound responsibility to work toward a better future that reflects God's ultimate purposes. Nor does he imply that the certain coming of the Son of Man means that his followers can sit around waiting for his return. Quite the opposite! Yet the work to which Jesus calls us is to be done out of reverence and love for our Lord. It involves entrusting ourselves to God in full confidence of his goodness and grace and in the future he has promised us: "Do not be afraid, little flock, for your Father has been pleased to give you the kingdom" (Lk 12:32).

8

Revelation and the Renewal of All Things

We stand with one hand on the door,
Looking into another world
That is this world, the pale daylight
Coming just as before, our chores
To do, the cattle all awake,
Our own white frozen breath hanging
In front of us; and we are here
As we have never been before,
Sighted as not before, our place
Holy, although we knew it not.

WENDELL BERRY,
"REMEMBERING THAT IT HAPPENED ONCE"

◆

HEAVENLY VISIONS, EARTHLY REALITIES

On a recent Sunday morning, the worship leader at the church one of us was attending read a passage from the book of Revelation and then urged the congregation to focus on the hope that we have to "escape earth and enter heaven." Such language may be so familiar to some of us that we rarely notice how unbiblical it is. This time, however, the dissonance was impossible to miss. Our reading had just ended with Revelation 5:9-10, a magnificent song of praise to Jesus:

> You are worthy to take the scroll
> > and to open its seals,
> because you were slain,
> > and with your blood you purchased for God
> > persons from every tribe and language and people and nation.
> You have made them to be a kingdom and priests to serve our God,
> > and they will reign on the earth.

Here as elsewhere in John's vision, Christ's atonement does not serve to open "escape hatches" for the redeemed to ascend to heaven; rather, Christ ransoms for God a people, a "priestly kingdom," who will reign on *earth*.[1] In the following verses, John's vision, like a camera slowly zooming out, expands from its focus on the throne and those gathered immediately around it to show how all of creation—"every creature in heaven and on earth and under the earth and on the sea, and all that is in them"—joins in the praise for God and the Lamb (Rev 5:13). As the rest of Revelation reveals, the future of all of creation has been secured in the victory of Christ, for now the "kingdom of the world" can become the "kingdom of our Lord and of his Messiah" (Rev 11:15). In the light of the death, resurrection and future return of the incarnate Christ, readers of John's Apocalypse are enabled to see this world through new eyes and to go about the work to which God calls us: to be here, as Wendell Berry's poem at the head of this chapter suggests,

> As we have never been before,
> Sighted as not before, our place
> Holy, although we knew it not.

The book of Revelation paints the fullest and grandest picture of Christian hope in Scripture. In his passionate, prophetic and pastoral letter to churches in Asia Minor—churches that are struggling in the late first century with everything from poverty, persecution and religious oppression to apathy and the temptation to give up and conform to their surrounding culture—John unveils the reality behind their situations and helps us gain a heavenly perspective on our own time too. Central to John's Apocalypse is the vision of God and the Lamb on the throne. No matter how chaotic the world looks or how despotic its earthly rulers, God in Christ remains the "ruler of the kings of the earth" (Rev 1:5) and his promises to his people and his purposes for his creation still stand. Revelation invites readers and hearers of this book to join in this praise that all of creation offers to the Creator and Redeemer, and it expands our imagination to consider what it looks like when God's will begins to be done "on earth as it is in heaven."

In its call to faithful witness and worship, and in its vision of this world transformed, John's Apocalypse enables us to discern the lies that seduce us into idolatry, selfishness and greed, and it makes impossible both complacency and despair. It is possible for us to join in the praise that all creation offers to God and the Lamb and to weep, even to groan; but we cannot despair when we see that the one we worship, though slain, stands alive and ready to give life. It is possible—and indeed right—for us to worship our Creator and Redeemer and to find peace and rest before his throne; but we will find it impossible to remain comfortable and at ease in the face of evil and injustice when we have once gazed on his beauty and holiness. Worship of God in Christ reorients us around our proper center and rules out the worship of anything else that would claim our affection.[2] By his light we are enabled to see light (Ps 36:9), and to see ourselves and our world from the perspective of eternity. Here is radical hope for any age, especially for an age prone to apathy, fear and despair.

That much of John's vision portrays upheaval, chaos and judgment should not surprise us. Such are the inevitable consequences of the encounter between God's righteousness and the forces of evil and injustice, the marks of a world that has set itself in opposition to God and his kingdom, the signs of the times when the powers that be seek to usurp the authority that belongs to God alone and to drag along a marveling populace behind them. John's vision invites us to see more clearly the true nature of this time between the times. He unveils for us the reality of this age of the "already and not yet." We live—to borrow Revelation's distinctive imagery—in light of victory already won: the Lamb that was slaughtered and yet lives opens the seals of the scroll that contains God's purposes for his creation (Rev 5:6-10), ensuring the realization of the promises of God; and Christ's victory means that Satan has been cast forever out of heaven, no longer able to accuse us before God's throne (Rev 12:10-11). But we also live in hope of what is yet to come: Satan has not yet been thrown into the lake of fire (Rev 20:10), God's people are still persecuted on earth (Rev 12:12-17), and the powers of destruction yet wreak havoc throughout God's creation. Even as we celebrate that "now have come the salvation and the power and the kingdom of our God, and the authority of his Messiah" (Rev 12:10), we long to see the ultimate realization of this reality in the coming of the New Jerusalem, in the new heaven and the new earth (Rev 21:1).

THE GLORY OF GOD AND THE FALL OF "BABYLON THE GREAT"

When the prophet Isaiah saw the throne of God, he heard seraphim crying out, "Holy, holy, holy is the LORD Almighty; the whole earth is full of his glory" (Is 6:3). In John's vision the four living creatures that surround God's throne cry out with a similar refrain, but with a distinctive difference: "Holy, holy, holy is the Lord God Almighty, who was, and is, and is to come" (Rev 4:8). In John's vision there is a keen awareness of how difficult it often is to discern the glory of God in an earth that seems hostage to the powers of destruction, in a world where the beasts of Revelation 13 hold sway and the idolatry of Babylon rules the day. Yet what John hears also reminds readers that God's very identity is as one who "comes" to his people. Not only has he always been, still is and always will be—the eternal God—but he remains intimately involved in his creation and will one day visit the world in judgment and redemption at the return of Christ.[3]

In this in-between time, John's vision challenges readers to resist the siren calls of our culture that would invite us to worship money, power and empire. This challenge is sometimes missed by present-day readers of Revelation 6–16, where John describes a vision that is populated with a dazzling array of images and stories drawn from Scripture and the ancient world. If twenty-first-century readers find these themes difficult to interpret without a lot of hard work, Revelation 17–19 brings some of them nearer to the surface and makes their import harder to miss—and as a result is likely to make contemporary readers squirm. The parallels to our own situation become a little too close for comfort. These chapters introduce us to "Babylon" (Rev 17), describe its fall, offer a series of laments over its fall (Rev 18) and give a counterpoint of praise to God and celebration of his judgment of Babylon (Rev 19). As readers, we are invited to consider where we stand. Are we so caught up in the life of Babylon that we would lament its fall? Or is our orientation toward God's kingdom such that we would find ourselves joining in the heavenly chorus of praise?

Somewhat unusually for Revelation, an angel explains to John (and to his readers) just what the Babylon of his vision represents. In biblical history, of course, Babylon was the nation that destroyed the Jerusalem temple and carried God's people into exile, so John and his first readers would no doubt already suspect that the mantle of Babylon could belong in the late first century only to the Roman Empire. Rome, for all that it claimed to bring peace and security to its conquered lands, bought its "peace" with war. In A.D. 66, Vespasian—soon to be emperor himself—crushed a revolt in Palestine, leading to the death of thousands of Jews. In A.D. 70, in the face of ongoing revolt, his son Titus (Vespasian was now emperor) finished the job, destroying the Jerusalem temple and carrying away its riches. The victory is commemorated in Rome with the Arch of Titus, adorned with scenes that visitors to the city can still see today, of captured Jews and temple implements being carried away by Roman soldiers. For Jews after A.D. 70, then, the connection between Rome and Babylon had become all too clear. And for Christians, whether Jewish or Gentile, the experience of living under the Roman Empire was often one of religious persecution, especially in Asia Minor where the emperor was venerated and worshiped as a god. For some, Nero (A.D. 55–68) had not been an aberration but rather was the very embodiment of the mad hubris of empire; and in Domitian, emperor from A.D. 81 to 96, the period in which Revelation was most likely written, some feared another Nero. If the historian Suetonius is to be believed, Domitian did not help allay anxiety about his character when he referred to himself as "our Lord and God" and expected others to address him as such (*Life of Domitian*, 13.2). John makes certain that readers cannot miss the association between Babylon and Rome: after seeing a prostitute, who seems to be a parody of the goddess Roma, sitting on a beast with seven heads, John learns in Revelation 17:9 that the seven heads of the beast on which Babylon sits are "seven hills."

Figure 8.1. The Arch of Titus in Rome. The Roman Emperor Domitian had this arch built in A.D. 82 to commemorate the victories of his recently deceased brother, the Emperor Titus. Among the victories commemorated is the destruction of Jerusalem and the carrying away of Jews and implements from the temple. In John's vision, Rome represents a new Babylon in its imperial power, violence, idolatrous trade and oppression of God's people.

Rome, of course, is famously built on seven hills, and no ancient reader of John's Apocalypse could fail to miss the connection.

What is portrayed in Revelation 17–18, then, is the fall of Rome as the embodiment of Babylon, of an empire that has set itself up in opposition to God and his people. We may be grateful today not to live in such an empire; most readers of this book will be members of societies in which there is substantial political and religious freedom, where there is little explicit interest in building the sort of political empires that depend on colonization and conquest, and where there is more equitable distribution of wealth and opportunity than there ever was in the Roman Empire (however much we rightly lament the substantial inequalities that remain in our own societies and across the world). Nonetheless, when we read the laments over Babylon's fall in Revelation 18, it is difficult not to see ourselves in the kings, the merchants and the ship captains who have invested all they have in Babylon's glittering promises of power, wealth and luxury. We may not live in Rome, but our consumerist societies can look an awful lot like Babylon.

Babylon claims to reign as queen (Rev 18:7), to be sovereign over the earth and its peoples; she promises worldly power to those who consort with her and untold luxury for those who participate in her trade (Rev 18:3). Her merchants, we learn, "were the world's important people" (Rev 18:23), and those who shipped her goods became rich through her wealth (Rev 18:19). Rome's imperial dominance and worldwide trade means that the goods listed in Revelation 18:11-13—which made the merchants of "Babylon" wealthy—are drawn from the remotest parts of the known world of the time. These items would be well known to John's first readers, even if none of them could ever have afforded the more expensive things on the list: there is Spanish gold and silver, African or Indian ivory (African elephants had already been hunted to such an extent that new sources of ivory were being sought in the East), Chinese silk, Arabian and Indian spices, olive oil and wine from throughout the empire.[4]

John also lists one final commodity that was similarly bought and sold everywhere: "human beings sold as slaves," or, as it might also be translated, "the bodies and souls of human beings" (Rev 18:13). This ought to jolt John's readers awake. However much we understandably want to celebrate the fruits of the earth and the riches of human culture and ingenuity represented in the rest of the list, this last item startles us. It forces us to confront the true cost of an economy that provides comfort and luxury for the few at the expense of the many; it calls attention to the dehumanizing consequences of directing worship toward anyone or anything that is not our true Creator and Lord. If we miss the point here, the end of the chapter goes on to reveal that in Babylon "was found the blood of prophets and of God's holy people, of all who have been slaughtered on the earth" (Rev 18:24). Indeed, the earth itself, we learn in Revelation 19:2, has been "corrupted" by Babylon's "adulteries" (a traditional biblical way of referring to the worship of gods other than the one God of Israel). These, then, are the true marks of Babylon: slavery, bloodshed, pollution of the earth. Yet why is this not perceived by those who lament over its fall?

The problem is that the "nations" are entranced by Babylon's beauty, power and wealth; they cannot see, or do not want to see, the price at which such wealth has been obtained or the death and destruction that the exercise of such power has caused. There is something "magical" about the lure of Babylon (Rev 18:23). John himself, and John's readers, are not immune to the spell. When John first encounters Babylon, in the form of a richly dressed royal prostitute (Rev 17:3-6), he says he "was greatly astonished" (Rev 17:6). The Greek word for "astonished" here is the same word that is used elsewhere in Revelation for "marveling," for an astonishment that includes wonder and even attraction. Those who are "astonished" at the beast in Revelation 13:3 marvel "after it," apparently following in its wake, and they worship the dragon—Satan himself—who lends his power to the beast (Rev 13:4). So John's "astonishment" when he sees Babylon the Great (who, we are told, is drunk with the blood of God's people) ought to concern readers. We are not surprised that the angel immediately asks John, rather pointedly, *why* he is astonished or marveling at Babylon. The angel says that he will explain the "mystery" of Babylon; the angel will break the spell that has momentarily captured John by revealing the true reality behind her appearances.

If John recognizes in himself the temptation to buy into the lie of Babylon, he knows that his readers face the same temptation. Thus he tells us that he hears a voice, just as Babylon is being judged, crying out with the warning, "'Come out of her, my people,' so that you will not share in her sins" (Rev 18:4). God's people do not live somewhere other than Babylon; in this age it is the place of their exile. But they are challenged nonetheless to "come out" from all that Babylon represents, to demonstrate that their citizenship is elsewhere. God's people are not immune to Babylon's charms: like anyone else, they are at risk of finding themselves enmeshed in the worship of power, wealth and luxury that defines the culture and society in which they live. The challenge for John's readers is the same as it was for those who heard Jeremiah when he issued the same call to the exiles in the original Babylon (Jer 51:45): to refuse to worship the idols of the empire and keep from participating in her sins and falling under the judgment that must surely befall her.

These chapters unmask the Babylon of John's day, and in our own age too they can prompt prophetic critique of those elements of our governments and societies that are Babylon-like. They ought above all to lead us to examine the orientation of our own lives, embedded as we are in consumerist cultures that depend so often on the exploitation of people and their lands and the ruin of the earth. We are liable to worship and invest our entire lives in the pursuit of the ephemeral promises of wealth and power. We are easily distracted and blinded from seeing the destruction of the earth and the exploitation of the poor that often accompanies such idolatry, all the more so when the worst effects of our heedless consumption and indulgent lifestyles are felt most in distant parts of the world. We rarely lament over a warming and abused planet where whole ecosystems, species and human communities are struggling—and sometimes failing—just to survive, where there are still "bodies and souls of human beings" sold as slaves to produce the goods we buy, and where the earth continues to be polluted and destroyed to sustain our way of life.

Yet we *would* lament if we saw our version of Babylon fall. So John has to unveil the reality of Babylon for us. He also, as we will see, has to give us an alternative vision in which to invest ourselves, a vision of God's creation as its Creator and Redeemer intends it to finally be.

The "Hallelujah Chorus"

For those who enjoy Handel's *Messiah,* it can come as a surprise to learn that the lines of the famous "Hallelujah Chorus" are drawn from a context of celebration over the judgment of Babylon. Yet all the "heavens" and the people of God are called on to rejoice over God's judgment of Babylon in Revelation 18:20, and Revelation 19 portrays their praise with a threefold "Hallelujah," the last of which shows up in the *Messiah*:

> Hallelujah!
> Salvation and glory and power belong to our God,
>   for true and just are his judgments.
> He has condemned the great prostitute
>   who corrupted the earth by her adulteries.
> He has avenged on her the blood of his servants. (Rev 19:1-2)
>
> Hallelujah!
>   The smoke from her [Babylon] goes up for ever and ever. (Rev 19:3)
>
> Hallelujah!
>   For our Lord God Almighty reigns. (Rev 19:6)

God's judgment of Babylon is integrally tied in Revelation to the salvation of his people and his creation. The collapse of Babylon, as the embodiment of a world order that sets itself up in the place of God, is the necessary consequence of the in-breaking of God's kingdom. John's readers are invited to inscribe themselves into the narrative here, to reject the temptation to join the kings and merchants who see in Babylon's fall only the loss of their power and wealth, and instead to join the "great multitude in heaven" (Rev 19:1), the representatives of the entire created order (Rev 19:4) and all of God's people (Rev 19:4, 6) in the praise of God that resounds in the heavenly throne room.

Revelation 19:1-10 picks up themes that have already appeared in Revelation 11:15-18, where the first half of Revelation comes to a close with a celebration of the coming of God's kingdom. In both texts we find God's "servants," both the "small" and the "great" (11:18; 19:5), praising God for his power (11:17; 19:1), his rule (11:15, 17; 19:6) and his judgment—in chapter 19 specifically of Babylon, and in chapter 11 of all those who "destroy the earth" (11:18). Readers of Revelation understandably may wonder about God's purposes for creation when so much of the book is taken up with scenes of destruction and cosmic upheaval, but Revelation 11:18 reminds us that God's intention is ultimately to reclaim his creation for himself. The time is coming for "destroying those who destroy the earth" (Rev 11:18).[5] The upheavals in creation are seen in this light as signs of the disorder that results from human idolatry and injustice, signs of

God using creation itself to fight back against the powers that are corrupting the earth and leading astray its peoples. It is significant that, at the opening of the seals of the scroll in Revelation 6, it is the four living creatures—representatives of all of creation[6]—who call forth the four horsemen of the Apocalypse. And if we want to see what God's ultimate purposes are for creation, Revelation invites us into a vision of a new heaven and a new earth, into the future that Christ's death and resurrection have secured.

A NEW EARTH

In Handel's "Hallelujah Chorus" the text of Revelation 19:6 is brought together with Revelation 11:15, which is an appropriate combination given the close connections we have observed between these passages. In Revelation 11:15 the seventh and final trumpet is blown and loud voices in heaven proclaim, "The kingdom of the world has become the kingdom of our Lord and of his Messiah, and he will reign for ever and ever." This transfer of authority occurs definitively at the end of the age, when "the time has come for judging the dead" and "destroying those who destroy the earth" (Rev 11:18), though it is anticipated already in Christ's death and resurrection. In Revelation 12, for example, a heavenly voice proclaims, after a symbolic portrayal of Jesus' resurrection and ascension (Rev 12:5), that "Now have come the salvation and the power and the kingdom of our God, and the authority of his Messiah" (Rev 12:10). So although it will not be seen in its fullness until the return of Christ, John is encouraging and challenging us to recognize that God's kingdom has already come and that the proper center of our authority and worship is God and his Christ (as we have seen in Rev 4–5). Returning to Revelation 11:15, it is striking that in the Greek text the word for "kingdom" (*basileia*) is not repeated, as in our English translations, resulting—as theologian Poul F. Guttesen points out—in the impression that it is the same entity, the same realm that is transferred from one ruler (Babylon and all it represents, we might say) to another (God and his Messiah).[7] Through this "regime change," as "God assumes this position that is rightfully his,"[8] the world of God's creation and human society is not abandoned but rather is reordered around its proper center. The picture is of this world wrested back from the powers that are destroying it and finally placed firmly under the reign of God (Rev 11:17).

What Revelation 11 describes in a few verses is developed at much greater length in Revelation 21–22. Here is where John, echoing Isaiah and Ezekiel, sees "a new heaven and a new earth" (Rev 21:1) and "the Holy City, the new Jerusalem, coming down out of heaven from God" (Rev 21:2). In this vision of humanity and earth renewed, of a world where there is no need for a temple because God's presence and glory suffuses every corner of it, we are invited through our imagination to be transformed by John's vision of creation's future and to reassess our present in its light. Central to the vision is God's intimate presence: "Look! God's dwelling place is now among the people, and he will dwell with them" (21:3). As Isaiah foresaw, the Lord himself will "wipe every tear from their eyes" (21:4; cf. Is 25:8). The new Jerusalem has no temple because "the Lord God Almighty and the Lamb are its temple" (Rev 21:22); and the entire city is described as a perfect cube (21:16), suggesting that all of creation has become like the Holy of Holies, full of the glorious presence of God. In the city are the throne of God and the Lamb (Rev 22:3), and his servants will see his face (22:4) and live in the light of his presence, reigning for ever and ever (22:5). Just as was anticipated by the worshipers in Revelation 5:10, those bought by the blood of the Lamb are now enabled to do what God always intended for us, to "reign on earth for ever and ever."

It is through God's redemptive, reconciling work in Christ that "everything" is made new (Rev 21:5), and this apparently includes all that is best in human culture, as the "kings of the earth" bring their "splendor" through its ever-open gates (Rev 21:24). Babylon entices us with its promises of riches, power, comfort and beauty, but here John hints at an alternative vision of human kingdoms and cultures renewed. He leaves it up to us to work out how we celebrate and enjoy all that God gives us in creation and in human ingenuity without turning his good gifts into idols that lead to destruction and ruin.

John's vision is not only of a city but of a whole new creation, a heaven and an earth, and in this cosmic vision "nature" and civilization are imagined together as what they can finally be only by the grace of God. Revelation 22 thus shows us the "river... of life" flowing through the city, around which grows the tree of life, bearing fruit every month and growing leaves that serve for the "healing of the nations" (Rev 22:2). In this Edenic vision, with its tree of life, we are reminded again of Genesis, as we were already by the mention of a "heaven" and an "earth." Here in the new heaven and new earth, the entire creation reaches the goal that God always intended for it. This is why God is described earlier in this text not as making *all new things*, but as making *all things* (or "everything") *new* (Rev 21:5). There is a fundamental continuity between this creation and the new creation, a continuity that gives us hope for this world in God's future and challenges us to anticipate his kingdom even in how we live and care for the earth now.

There must of course be a radical discontinuity if the kingdom of this world is to become the kingdom of the Lord and his Messiah, and John tries to help us understand the nature of this break by listing seven things that are "no longer." The absence of these seven things—not a coincidental number in Revelation—reveals to us what it means for the "first heaven and the first earth" to have "passed away" (Rev 21:1). The seven things that John describes as being "no longer" (Gk. *ouk... eti*) are as follows:

> the sea (21:1)
>   death (21:4)
>   mourning (21:4)
>   crying (21:4)
>   pain (21:4)
>   curse (22:3)
> night (22:5)

The central three items on this list, mourning, crying and pain, are all marks of a world held hostage to death and plagued by its curse. They are no more when death itself is swallowed up and "any curse" is removed. John is alluding again to the Genesis narrative, this time to Genesis 3, portraying a world in which the results of humankind's rebellion against God have been reversed. In the new creation, the curse and the death that mark life in this age are gone and the conditions that were meant to apply in the Garden of Eden are restored. Yet John's vision of a new creation is not merely of a world in which the clock has been turned back to its beginnings; it is rather of a world taken forward into the future God always intended for it.

This is probably why the new heaven and new earth are described as having no "sea" or "night." The absence of the sea in particular often puzzles readers of Revelation. John's vision, after all, includes the sea in the praises of God and the Lamb (Rev 5:13) and pictures it as part of God's good creation (Rev 10:6; 14:7), so he clearly has nothing against the ocean in a literal sense. John may even intend readers to empathize with the plight of the creatures in the sea when, in the midst of the judgments described in Revelation 8:9, he says that a third of the creatures in the sea died—"the ones having souls," as it might be translated (Rev 8:9; cf. 16:3). The word John uses, *psychē*, is indeed the Greek word for "soul," though here as in other biblical contexts it refers not to soul as a separable or immaterial part of one's being but to the very "life" or "breath" of these creatures, perhaps with an echo of God's life-giving breath or Spirit that makes them (and us) into "living beings" (as in Gen 1:20-27; 2:7). In any case, John's description may serve to call attention to the significance of the death of these creatures that have been given their life by God. But if the literal ocean can be a benign part of God's creation for John, the "sea" of his biblically informed imagination represents something more: the threat of chaos and the judgment of God.

As in the visions of Daniel 7, the mythological sea is the source of a terrifying beast in Revelation 13:1 and it is linked with the "abyss" itself (Rev 11:7; 17:8). It is also used to symbolize the instrument of judgment used to destroy Babylon, which is cast like a millstone into the sea (Rev 18:21), a fitting punishment for an empire that made its living on the sea. The sea is linked in the Old Testament to terrifying beasts like the Leviathan, to the threat of chaos and to God's judgment, whether in Noah's flood or the destruction of the Egyptians in the sea at the exodus. The darkness of night too can hint at danger and represent the judgment of God (Rev 16:10). So the doing away with *this* sea and the darkness of night means for John's readers the removal of all threat of judgment, of all potential for evil to arise in the new creation.

In Genesis the "waters" and "darkness" (Gen 1:2) that are present before God says "Let there be light" could hint at the potential threat present even in God's good creation. Though no threat to God, who tells the waters where to stop (Job 38:11) and for whom even Leviathan could be a source of delight (Ps 104:26), these waters and darkness nonetheless represent a latent threat to human life and security—a threat realized in the rebellion of human beings against their Creator. A broken relationship with God meant that our relationship with creation too was broken, and so the terror of the mythological sea and night became reality, and the threat of judgment hung over us. But John's vision suggests that the new creation is taken beyond all such threat; there is no risk of reversion to evil, chaos and judgment.[9] The new heaven and new earth are not simply a recapitulation of the past: they represent this heaven and earth brought to their divinely intended *telos*, made possible only by what God has done in Christ.[10]

WONDER, PRAISE AND THE IMAGINATION

John's need to resort to descriptions of the new creation *via negativa*, by emphasizing what is *not* there, suggests that the only way our imagination can begin to grasp God's purposes for the future of creation is by starting with what we have here and now. The world to come can only be envisioned as this world taken beyond all threat of evil and destruction. The promise of the new heaven and new earth is that "everything" is made new, and so we are freed to embrace the beauty and wonder of this creation and the joy and love that God refreshes us with even in the midst of this life. We are not ignorant of the brokenness of this age; we are rightly wary of the temptation to turn even God's good gifts into idols; and we must not mistake our proximate hopes for seeing God's kingdom instantiated in the here and now with the ultimate hope of what God has promised one day to bring about.[11] Yet for all that, John's vision invites us to see God at work even in our humble efforts; to sense the weight of his presence in the midst of our suffering, faithful witness; and to perceive in the stabs of joy we are sometimes granted a foretaste of the fullness of life only God in Christ can provide. We are invited most of all to join in the praise that all the rest of God's creatures offer him merely by being themselves; to see our place again as holy, though we knew it not; and to offer ourselves as living sacrifices in the service of our King, the only logical response and means of worship possible for us.

The despair that tempts us when the world looks far from reaching God's purposes, when we see a creation that is subjected to futility and handed over to the destroyers of the earth, is a risk that John too faced, even in the very throne room of God. When he first sees the scroll in God's hand, the scroll that represents God's purposes for his creation, it seems for a moment that they will never be fulfilled. There is no one, after all, who is worthy to break the scroll's seals, to open it or even to look in it (Rev 5:4). John's response is the only possible response to a situation in which the world as it is, with its death, destruction and sorrow, will be left to fall apart; where the earth and all its life will be given over to ruin; where there is no hope. He weeps. So too might we weep, not least when we recognize in our own time the plight of the earth and so many of its creatures. But this is not the end of the story.

It turns out that there is one who is worthy to open the scroll. The Lion of Judah, become a slain Lamb, is by his blood reconciling the world to God. Now God's people are made able to fulfill God's purposes: as a kingdom and priests, "they will reign on the earth" (Rev 5:10). There is confusion in the manuscript tradition for this verse about whether the reign is present or future, but such ambiguity is the hallmark of John's Apocalypse—and of the whole New Testament. The future has broken into the present in Christ, and so the promise of new creation and our reign as God's image bearers—reconciled to God, to each other and to creation too—begins to be realized now. As Revelation and the Psalms remind us, it is above all in worship that we are enabled again to see our present in the light of God's future, to discover afresh our proper place within God's creation, and to find ourselves transformed by the renewing of our minds and hearts so that we can persevere in bearing faithful witness to the Lamb.

9

Finding Joy in an Active and Living Hope

The Lord is the everlasting God,
the Creator of the ends of the earth.
He will not grow tired or weary,
and his understanding no one can fathom.
He gives strength to the weary
and increases the power of the weak.
Even youths grow tired and weary,
and young men stumble and fall;
but those who hope in the Lord
will renew their strength.
They will soar on wings like eagles;
they will run and not grow weary,
they will walk and not be faint.

Isaiah 40:28-31

A Rocha, and the rapidly growing number of those like us around the wide country of the USA, will never promise to "save the earth." Just to see the signs of the kingdom of God written in restored landscapes, healed communities and lives renewed in the Holy Spirit who breathes on the earth.

Peter Harris, founder of A Rocha,
a Christian environmental charity

◆

Not of This World?

In the Gospel of John, Jesus tells Pilate that his kingdom is "not from this world" (Jn 18:36, our translation). The expression has often been translated "my kingdom is not *of* this world," leading some to conclude that Jesus' kingly rule does not have much to do with the messy business of life here on earth. But Jesus' point here is that his kingdom does not have its *origin* in this world, nor is it arranged in the same way as the world's kingdoms, which depend on violence of the sort Pilate might otherwise have expected from Jesus' followers. Jesus' kingdom is not *from* this world, but it most assuredly is *for* this world. He came to seek and to save the lost, to inaugurate the kingdom of God, to atone for the sins of the world on the cross, to redeem and reclaim a suffering creation and to prepare the way for the renewal of all things—of which his physical resurrection from the dead is the sure and certain sign. In this time between his resurrection and return, Jesus' followers have work to do.

Jesus gives us a good idea of what this work looks like when we entrust ourselves to God and live in lively expectation of his return. When at the end of Matthew's Gospel he gives his disciples the Great Commission, Jesus holds together making disciples of all nations (which of course requires the proclamation of the gospel), baptizing them into the Father, Son and Holy Spirit, and teaching them to obey all that he has commanded. We have in this book explored one element of what such obedience to Christ as Lord looks like, of what it means to be a disciple of Christ today in our context. We have considered the implications of God's gift of the kingdom for how we care for his creation and prepare in hope for God's future in the midst of a world given over to doubt and despair.

Always Pray and Do Not Give Up

Above all, Jesus assures us that God has not abandoned us. He is with us always, "to the very end of the age" (Mt 28:20). He will not "keep putting off" all those who "cry out to him day and night" for justice (Lk 18:7). God's love in Christ for his people and for all his creation ensures that he will fulfill his promise to come to the earth to judge injustice and reclaim what is rightfully his. One day all things will be put to rights, and it is this hope that encourages and sustains all God's people who long to see his justice done, who long to see the earth become a place where "righteousness dwells." Jesus expects his followers to be persistent, to persevere and not to give up, like the widow in his parable who goes on begging even an unjust judge for justice. Luke tells us that Jesus told the parable of the widow precisely so that his disciples would "always pray and not give up" (Lk 18:1). The same challenge applies to us today whenever we are tempted to give in to despair.

Yet when we cry out in prayer for justice to be done, we cannot avoid also reflecting on the extent to which our own lives and actions are deserving of God's judgment. Do our own lives display God's justice and righteousness? As Jesus asks his disciples, "When the Son of Man comes, will he find faith on the earth?" (Lk 18:8). Will he find us faithfully proclaiming the gospel and caring for our human and nonhuman neighbors, for the world that belongs to him and for which he died so that all things might be reconciled to God?

No Excuses

We are all—including the authors of this book—tempted to find ways to avoid the radical implications of God's Word for how we live in creation today. One of the easiest ways to escape our responsibility is to plead ignorance. We simply do not know, we say, nor can we know, precisely how to go about reflecting God's love in our relationships to the nonhuman creation. After all, we rightly delight in the ways in which God provides for us through creation, eating what it produces, making buildings from its materials, creating beauty with its resources. Our relationship with nonhuman creation is obviously different in kind from our relationship with fellow human beings. So if there are plenty of difficult ethical questions that arise for Christians concerning interpersonal relations—questions on which Scripture is often unclear—then it may be all the more difficult to agree on what it means to live ethically in our relations with the nonhuman creation. Where are we to go for guidance in weighing trade-offs and in making the many detailed, difficult decisions that face individuals, churches and society as a whole?

We suggest that Scripture provides us not with all of the answers in a simple list of dos and don'ts, but with a countercultural vision of radical discipleship,[1] a godly wisdom and an ethos that fundamentally reorients us to the world. Scripture challenges us to recognize the intrinsic value of all of creation (intrinsic in the sense of given by God) and of the particular value of human beings within creation; and it calls on us to reflect Christ's sacrificial love in all of our relations. Given that creation has value before God, we rightly use the abilities that God has given us to study creation, including the gifts of science and technology, to help us make the best decisions we can in caring for it and for its human inhabitants. We need to seek wisdom, discernment and humility in all our study and decision making. We must, however, be wary of our predilection to find answers that suit our own selfish desires, for we know that our hearts are deceitful, that we are always tempted to avoid the radical challenges of following in the path of Christ.

We cannot adopt a purely consequentialist ethic that assumes any means are acceptable so long as they bring about the ends we think God desires. We recognize that some cures are out of bounds, that some ends do not justify the means. Above all, we must be prepared to sacrifice our own comfort and to reject the idols of our consumerist societies. We must begin to engage in the difficult tasks of transforming our exploitative relationships with the earth and with our neighbors into relationships that more nearly reflect those for which God intends us. The challenging results of embracing such an ethos in the context of what science is telling us about the state of the earth today may well frighten us; but as we have been reminded over and again, "perfect love drives out fear" (1 Jn 4:18). God does not leave us alone in our task.

The second way we may try to evade our calling to care for creation is related to the first: we might claim that we must care first and foremost for our human neighbors and not get caught up in misanthropic "green" concerns that seem to neglect those who bear the very image of God. This excuse is related to the first because it reflects what today can only be an argument grounded in an almost willful ignorance, a refusal to see that it is impossible to care for our human brothers and sisters without caring for the environment in which they live. More nuanced arguments are sometimes made along this line about, for example, the prioritization of human welfare over the preservation of biodiversity. Most often, however, such prioritization turns out to be rooted in false trade-offs that ignore longer-term consequences or that depend on conducting business as usual and hence are unimaginative in coming up with fresh and better ways forward. We also must remember, based on the testimony of Scripture, that nonhuman creation too has value before God, and so we are compelled wherever possible to come up with solutions that enhance the health of both human and nonhuman communities.

The third excuse that we are tempted to employ in order to avoid contemplating radical changes in our lives is the most pious—and possibly the most dangerous in its consequences for life on earth. This is the apparently humble claim that human beings cannot possibly matter very much in the big scheme of things. Many of those who dispute the human influence on climate change adopt this posture. After giving a public lecture on climate change one of us had a comment from a Christian that since carbon dioxide only makes up 0.03 percent of the atmosphere, how could even doubling that amount possibly make any difference at all? It is tempting to write off many decades of scientific work and insight when it seems so reasonable to question how we puny creatures can possibly have any effect on a system as enormous and complex as the planet's climate. This claim is especially appealing to Christians, because we recognize humility as an important, even central virtue. Indeed, humility must be one of the defining characteristics of Christian creation care. Our current ecological crises would not have arisen if we as a species had remembered our own limits, our own "creatureliness": that we are a part of God's creation, not gods ourselves.[2]

In the present case, however, humility can be used as a smokescreen to disguise willful ignorance and abrogation of responsibility, and it serves to allow us to go on living however we like, oblivious of the consequences. We are reminded of a headline in the satirical newspaper *The Onion*: "'How Bad for the Environment Can Throwing Away One Plastic Bottle Be?' 30 Million People Wonder."[3] As chapters 2 and 3 demonstrated, the best scientists and scientific bodies around the world no longer have any doubt that the collective actions of over 7 billion people with ever-increasing rates of consumption are having demonstrable, dramatic effects on all sorts of large-scale biological and physical processes on earth. Long before scientists began making these observations and suggested that we label this age the "Anthropocene," Scripture made it clear in any case that we bear much responsibility for God's creation and that its fate is, in the end, bound up with our own.

LOVE, HOPE AND JOY

Biblical hope is not a promise of "pie in the sky by and by." Even as it assures us of God's faithfulness and of the glorious future we and all of creation have in Christ, it challenges us to faithful, righteous living that embodies God's promises in the here and now. Scripture's warnings of judgment, of being prepared for the return of the Master and the coming of the Son of Man, remind us of the seriousness of this challenge. When Jesus talks of the coming of the Son of Man, his emphasis is on what that means for how we live *now* in expectation of that coming. Jesus' teaching about the kingdom helps us to see the kingdom of God as already present in our encounter with him and challenges us to consider the ways in which God's kingdom reconstructs all of life around God's priorities. Yet if the challenges of life in the kingdom apply now, so too are we encouraged that there are foretastes of the blessings of the kingdom even in this age.

When Jesus challenged a rich ruler to "sell everything you have and give to the poor, and you will have treasure in heaven," the man "became very sad, because he was very wealthy" (Lk 18:22-23). Here is the challenge of the kingdom, and Jesus acknowledges how difficult it is for the rich to enter God's kingdom (Lk 18:25). But Jesus' disciple Peter seems to be reduced nearly to tears by this scene. Having watched the rich man walk away, this man who appears to be so close to the kingdom in his law keeping and faithful obedience to God, Peter wonders whether there is any hope for anyone. Is it so impossible to enter the kingdom of God? In weariness, perhaps, in confusion, or maybe even near despair, Peter says to Jesus, "We have left all we had to follow you!" (Lk 18:28).[4] Has it been in vain that he and the other disciples have given up all else to follow Jesus? The demands of the kingdom seem too heavy, the rewards too remote and uncertain.

Jesus immediately reassures Peter, and all those who would follow him in the way of the cross, that "no one who has left home or wife or brothers or sisters or parents or children for the sake of the kingdom of God will fail to receive many times as much in this age, and in the age to come eternal life" (Lk 18:29-30). In the version of this story that we have in Matthew's Gospel, the promised reward is described as entirely future, to be obtained at the *palingenesia*, which we might translate as "new creation" or "renewal of all things" (Mt 19:28). But in Luke's version we are reminded that the blessings of the age to come begin even now. The future promised in Christ challenges us to faithful discipleship now. It also shines light and hope back into our present to sustain, refresh and renew us. To live life for Jesus' sake is truly to find it, now and in the age to come (see Mt 10:39; 16:25). In his service is found rest, refreshment, joy and life: "Come to me, all you who are weary and burdened, and I will give you rest. Take my yoke upon you and learn from me, for I am gentle and humble in heart, and you will find rest for your souls. For my yoke is easy and my burden is light" (Mt 11:28-30). In place of the crushing weight of work that finds its source of strength and its end only in ourselves, Jesus offers his followers life-giving grace and rest in the work of his kingdom.

Love and joy are therefore the marks of the radical hope of the gospel grounded in faith in Christ. Peter reminds us of the source of such joy when, near the beginning of the book of Acts, he calls his fellow Israelites to repentance: "Repent, then, and turn to God, so that your sins may be wiped out, that times of refreshing may come from the Lord, and that he may send the Messiah, who has been appointed for you—even Jesus. Heaven must receive him until the time comes for God to restore everything, as he promised long ago through his holy prophets" (Acts 3:19-21). Peter links the return of Christ to the Old Testament promises of a new heaven and a new earth, when God would restore "everything" and make "all things" new (cf. Rev 21:5). Yet even as we await the return of Christ from heaven, Peter suggests that for those whose sins have been wiped away by Christ, for those who have repented and turned to God, there are in this age "times of refreshing" that come from the Lord. God through his life-giving Spirit renews and refreshes us even now, giving us foretastes of the new creation to come, joining us to Christ, the source of our hope and joy.

Yet our joy in Christ does not preclude "groaning" and lament. In fact, it is often only in lament and groaning alongside all of creation, in concert with the Holy Spirit, that we are enabled to persevere and are kept from giving up.[5] We may not mourn or grieve as others mourn (1 Thess 4:13), in hopelessness and despair, but Christians do—and must—mourn. In mourning alongside our sisters and brothers, we reveal that we are joined together in the body of Christ (Rom 12:15). In groaning alongside a creation enslaved to ruin, we acknowledge its plight (and our part in it) even as we yearn for its liberation and renewal. Groaning is our honest response to a world out of joint, to the recognition that things are not as they ought to be—not as God finally intends them to be. Paul assumes that groaning marks the Christian's life as surely as joy. Jesus, when Lazarus died, knew that Lazarus would be raised on the last day (as Martha too recognized) and, more than this, that even in the very next moment he would call Lazarus out of his tomb, alive. Yet standing there outside the tomb, Jesus wept (Jn 11:35). Our sure and certain hope in the resurrection and the new creation does not keep us from weeping while we yet live in a world of wounds. Only in the new heaven and the new earth will our tears be wiped away, once and for all, by God himself.

In the midst of our tears and our groaning, in our hope and in our joy, through our worship, we have work to do. In the gospel of Jesus Christ, we are promised a future—a future for life on earth. Although we must not confuse the ultimate hope we have in Christ for what God will bring about in new creation with our own always insufficient attempts to reflect God's glory in our work,[6] let us nonetheless celebrate the foretastes of new creation that God grants us by his grace. The beauty and wonder of a world that even in the midst of threat testifies to his glory; the sight of a landscape that in its very wildness reminds us of our creatureliness and humbles us before the Creator; the beauty of land once degraded and apparently ruined, now restored and coming alive again (see figure 9.1); the joy and peace of human communities whose connections to the land and sources of well-being have been regained and renewed; and even the groaning in sorrow when we witness the opposite of all this and yearn for the world to become the place even our biblically inspired imagination can only glimpse—all of this becomes a part of our praise as we join in the cosmic chorus of worship that is offered unceasingly before God's throne as he reconciles all things to himself in Christ.

Figure 9.1. The Aammiq wetland in the Bekaa Valley of Lebanon before and after A Rocha Lebanon brought conservation efforts to bear on this internationally important site for migrating birds. (Photo: A Rocha Lebanon)

Afterword

Practical Resources

If you read history you will find that the Christians who did most for the present world were just those who thought most of the next. . . . Aim at heaven and you will get earth "thrown in": aim at earth and you will get neither.

C. S. Lewis, *Mere Christianity*

◆

The way we live our daily lives is a central part of our worship of God, more so indeed than the hour or two we may spend in church on a Sunday. We live in expectation of the return of Christ and the renewal of all things. This means that our daily decisions should be shaped by the priorities of the kingdom of heaven that Jesus announced, here already in a veiled and partial way, but the fullness of which is yet to come. That understanding ought to govern our attitudes toward the work to which God has called us in his creation.

The way in which we care for creation—and support others in their task—will depend on the context in which each one of us lives, so we do not wish to be prescriptive with a big list of dos and don'ts. But the following organizations provide useful resources, and it would be well worth getting involved with one or more of them.

A ROCHA

www.arocha.org

An international Christian organization that, "inspired by God's love, engages in scientific research, environmental education and community-based conservation projects." Their US-based website is www.arocha-usa.org.

CARE OF CREATION

www.careofcreation.net

Care of Creation's core objectives are about pursuing a God-centered response to the environmental crisis. They think that the people who believe God made the world should be passionate about taking care of it.

EVANGELICAL ENVIRONMENTAL NETWORK

www.creationcare.org

The Evangelical Environmental Network "seeks to equip, inspire, disciple and mobilize God's people in their effort to care for God's creation." They believe "the body of Christ should be an example of what God's people can do in the world to solve some of the great challenges of our time."

RESTORING EDEN

www.restoringeden.org

"Restoring Eden is a movement of like-minded people who see a strong connection between our Chritian spirituality and our role as caretakers of creation." Their mission is "to make hearts bigger, hands dirtier, and voices stronger by rediscovering the biblical call to love, serve and protect God's creation."

RENEWAL

www.renewingcreation.org

"Renewal is a Christ-centered and student-driven creation care network that strives to inspire, connect and equip college students in their sustainability efforts."

Practical Resources

THE CENTER FOR ENVIRONMENTAL LEADERSHIP
www.center4eleadership.org
The Center for Environmental Leadership is "devoted to two things: helping individuals, institutions and communities act on their convictions to care for creation, and educating the next generation of Christian environmental leaders."

YOUNG EVANGELICALS FOR CLIMATE ACTION
www.yecaction.org
"Young evangelicals in the United States who are coming together and taking action to overcome the climate crisis as part of our Christian discipleship and witness."

AU SABLE INSTITUTE
www.ausable.org
"Au Sable Institute of Environmental Studies offers environmental science programs for students and adults of all ages: primary and secondary school, college, and graduate school."

FARADAY INSTITUTE FOR SCIENCE AND RELIGION
www.faraday-institute.org
The website contains many resources on science and religion, including a number of papers and lectures on Christianity and the environment that are all free to download.

Notes

1. Apocalypse Now? Living in the Last Days

[1] Jon Mooallem, "The End Is Near (Yay!)," *The New York Times Magazine*, April 16, 2009.
[2] A list of projects and initiatives linked to the Transition Network is maintained at www.transitionNetwork.org.
[3] Martin Rees, *Our Final Century* (Arrow Books, 2006).
[4] George Monbiot, "Is There Any Point in Fighting to Stave Off Industrial Apocalypse?" *The Guardian*, August 17, 2009, www.guardian.co.uk/commentisfree/cif-green/2009/aug/17/environment-climate-change.
[5] Stephen Mulvey, "Averting a Perfect Storm of Shortages," BBC News, August 24, 2009, http://news.bbc.co.uk/1/hi/sci/tech/8213884.stm.
[6] Samantha Weinberg, "The Natural-History Man," *Intelligent Life*, September/October 2012, http://moreintelligentlife.com/content/ideas/history-a-natural.
[7] "PM Warns of Climate Catastrophe," BBC News, October 19, 2009, http://news.bbc.co.uk/1/hi/uk/8313672.stm.
[8] Slavoj Žižek, *Living in the End Times* (New York: Verso, 2010).
[9] See Kevin Maher, "Apocalypse Everywhere," *Times Online*, November 13, 2009, http://entertainment.timesonline.co.uk/tol/arts_and_entertainment/film/article6914402.ece; Toby Lichtig, "Apocalypse Literature Now, and Then," *The Guardian*, January 20, 2010, www.guardian.co.uk/books/booksblog/2010/jan/20/apocalypse-literature-now.
[10] Editorial summary in *Nature* 461 (2009): 461, 472-75.
[11] See James J. McCarthy, "Climate Science and Its Distortion and Denial by the Misinformation Industry," in Robert S. White, ed., *Creation in Crisis* (London: SPCK, 2009), pp. 34-52; James Hoggan and Richard Littlemore, *Climate Cover-Up* (Vancouver: Greystone Books, 2009).
[12] Alan Weisman, *The World Without Us* (New York: Picador, 2007).

2. Life on Earth Today

[1] The superscription of Psalm 8 associates David with this psalm, and for convenience we here refer to him as the author.
[2] Carl Sagan, *Pale Blue Dot: A Vision of the Human Future in Space* (New York: Random House, 1994), p. 7.
[3] Mathis Wackernagel and William Rees, *Our Ecological Footprint: Reducing Human Impact on the Earth* (Gabriola Island, BC: New Society Publishers, 1995). See also www.footprintnetwork.org for up-to-date calculations of our ecological footprint.

⁴An explanation of slow, transitional cultural and social changes is described by Jan Boersema, *Beelden van Paaseiland. Over de duurzaamheid van een cultuur* (*Statues of Easter Island, On the Sustainability of a Culture*) (Amsterdam: Uitgeverij Atlas, 2011). A classic example of the theory of catastrophic collapse as a result of environmental exploitation is presented by Jared Diamond in *Collapse: How Societies Choose to Fail or Survive* (New York: Viking Penguin, 2005).
⁵See www.millennium-project.org for details.
⁶United Nations, Department of Economic and Social Affairs, Population Division, *World Urbanization Prospects: The 2011 Revision*, ST/ESA/SER.A/322 (United Nations, 2012).
⁷Christians in rich countries might also consider having fewer of their own children in the light of the New Testament emphasis on how family is redefined and expanded to include all of those in Christ.
⁸United Nations, *World Urbanization Prospects: The 2011 Revision*.
⁹Thomas Malthus, *An Essay on the Principle of Population* (printed for J. Johnson, in St Paul's Church-Yard, 1798), chap. 7.20.
¹⁰Paul R. Ehrlich, *The Population Bomb* (Buccaneer Books, 1968), p. xi. Ehrlich has more recently claimed that there is now only a 10 percent chance of avoiding a collapse of global civilization; see Juliette Jowit, "Paul Ehrlich, A Prophet of Global Population Doom Who Is Gloomier Than Ever," *The Guardian*, October 23, 2011, www.guardian.co.uk/environment/2011/oct/23/paul-ehrlich-global-collapse-warning.
¹¹Mark Lynas, *The God Species: How the Planet Can Survive the Age of Humans* (Washington, DC: National Geographic Society, 2011).
¹²Diamond, *Collapse*.
¹³Joel E. Cohen, *How Many People Can the Earth Support?* (New York: W. W. Norton & Co., 1995).
¹⁴The Millennium Ecosystem Assessment reports are available free at www.maweb.org.
¹⁵For IPCC reports see www.ipcc.ch.
¹⁶See www.footprintnetwork.org.
¹⁷See for example James Lovelock, *The Revenge of Gaia: Why the Earth Is Fighting Back and How We Can Still Save Humanity* (London: Allen Lane, 2006).
¹⁸See Johan Rockström et al., "A Safe Operating Space for Humanity," *Nature* 461 (2009): 472-75, available with commentaries and full report at http://tinyurl.com/planetboundaries.
¹⁹ Whole Systems Foundation, "Species Extinction and Human Population," www.whole-systems.org/extinctions.html.
²⁰Lynas, *The God Species*, p. 32.
²¹Anthony Barnosky et al., "Has the Earth's Sixth Mass Extinction Already Arrived?" *Nature* 471 (2011): 51-57.
²²David Butler and Don Merton, *The Black Robin* (Oxford: Oxford University Press, 1992).
²³Aldo Leopold, *A Sand County Almanac* (New York: Ballantine Books, 1966), p. 190.

[24] S. G. Potts et al., "Global Pollinator Declines; Trends, Impacts and Drivers," *Trends in Ecology and Evolution* 25 (2010): 345-53.

[25] See Tom D. Breese, Stuart P. M. Roberts and Simon G. Potts, *The Decline of England's Bees* (London: Friends of the Earth, 2012), www.foe.co.uk/beesreport, and references therein.

[26] Robert Costanza et al., "The Value of the World's Ecosystem Services and Natural Capital," *Nature* 387 (1997): 253-60.

[27] Theodore Roosevelt, "Seventh Annual Message to the Senate and House of Representatives," December 3, 1907, available at www.presidency.ucsb.edu/ws/index.php?pid=29548.

[28] Aldo Leopold, *Round River* (Oxford: Oxford University Press, 1972), p. 165.

[29] Edward O. Wilson, *The Creation: An Appeal to Save Life on Earth* (W. W. Norton & Company, 2006).

[30] Natasha Gilbert, "Water Under Pressure," *Nature* 257 (2012): 256-57.

[31] An estimate of 47% is given in the *OECD Environment Outlook to 2030* (Organization for Economic Co-operation and Development, 2008).

[32] In a series of haunting, deeply troubling photographs, the artist Chris Jordan has documented the impact of plastic rubbish on albatrosses on Midway Atoll in the middle of the Pacific Ocean, 2,000 miles from the nearest continent. See http://chrisjordan.com/gallery/midway.

[33] S. C. Doney et al., "Ocean Acidification: The Other CO_2 Problem," *Annual Reviews of Marine Science* 1 (2009): 169-92.

[34] H. W. Kendall and D. Pimentel, "Constraints on the Expansion of the Global Food Supply," *Ambio* 23 (1994): 198-205.

[35] N. Gruber and J. N. Galloway, "An Earth System Perspective of the Global Nitrogen Cycle," *Nature* 451 (2008): 293-96.

[36] See Ellen F. Davis, "Just Food: A Biblical Perspective on Culture and Agriculture," in Robert S. White, ed., *Creation in Crisis* (London: SPCK, 2009), pp. 122-36; Davis, *Scripture, Culture, and Agriculture: An Agrarian Reading of the Bible* (Cambridge: Cambridge University Press, 2008).

[37] D. A. Pfeiffer, *Eating Fossil Fuels: Oil, Food and the Coming Crisis in Agriculture* (Gabriola Island, BC: New Society Publishers, 2006).

[38] Jenny Gustavsson et al., *Global Food Losses and Food Waste: Extent, Causes and Prevention* (Rome: Food and Agriculture Organization of the United Nations, 2011), www.fao.org/docrep/014/mb060e/mb060e00.pdf.

[39] Data from United States Department of Agriculture Economic Research Service. See www.ers.usda.gov/Briefing/CPIFoodAndExpenditures/Data/table7.htm.

[40] This statistic is reported by OECD Insights (available at http://oecdinsights.org/2010/01/25/biofuel), based on a study by the Earth Policy Institute.

[41] L. T. Evans, *Feeding the Ten Billion: Plants and Population Growth* (Cambridge: Cambridge University Press, 1998).

[42] The figures for the effect of livestock discussed in this chapter come from a compre-

hensive study by The Livestock, Environment and Development (LEAD) Initiative: *Livestock's Long Shadow: Environmental Issues and Options* (Rome: Food and Agriculture Organization of the United Nations, 2006).

[43]Ibid.

[44]See Calvin B. DeWitt, "Unsustainable Agriculture and Land Use: Restoring Stewardship for Biospheric Sustainability," in Robert S. White, ed., *Creation in Crisis* (London: SPCK, 2009), pp. 137-56.

[45]This data is based on that given by the Brazilian National Institute of Space Research and is available at http://rainforests.mongabay.com/amazon/deforestation_calculations.html.

[46]See the report of the Environmental Protection Agency, "Regulation of Hydraulic Fracturing Under the Safe Drinking Water Act," http://water.epa.gov/type/groundwater/uic/class2/hydraulicfracturing/wells_hydroreg.cfm.

[47]Michelle Bamberger and Robert E. Oswald, "Impacts of Gas Drilling on Human and Animal Health," *New Solutions* 22 (2012): 51-77, quoted from p. 51.

[48]See, e.g., William J. Ripple and Robert L. Beschta, "Wolf Reintroduction, Predation Risk, and Cottonwood Recovery in Yellowstone National Park," *Forest Ecology and Management* 184 (2003): 299-313; Ripple and Beschta, "Trophic Cascades in Yellowstone: The First 15 Years After Wolf Reintroduction," *Biological Conservation* 145 (2012): 205-13.

[49]For details, see Stuart Chape, Mark Spalding and Martin Jenkins, *The World's Protected Areas: Status, Values and Prospects in the 21st Century* (UNEP-WCMC and University of California Press, 2008).

[50]Norman Myers et al., "Biodiversity Hotspots for Conservation Priorities," *Nature* 403 (2000): 853-58.

[51]Hilary Marlow, *The Earth Is the Lord's: A Biblical Response to Environmental Issues* (Cambridge: Grove Books, 2008).

3. GLOBAL CLIMATE CHANGE

[1]United Nations Framework Convention on Climate Change, *Report of the Conference of the Parties on Its Fifteenth Session, Held in Copenhagen from 7 to 19 December 2009, Addendum Part Two: Action Taken by the Conference of the Parties at Its Fifteenth Session* (United Nations, 2010).

[2]Joeri Rogelj et al., "Copenhagen Accord Pledges Are Paltry," *Nature* 464 (2010): 1126-28.

[3]Eric Klinenberg, *Heat Wave: A Social Autopsy of Disaster in Chicago, Il* (Chicago University Press, 2002).

[4]Christoph Schär and Gerd Jendritzky, "Hot News from Summer 2003," *Nature* 432 (2004): 559-60; P. A. Stott, D. A. Stone and M. R. Allen, "Human Contribution to the European Heatwave of 2003," *Nature* 432 (2004): 610-14.

[5]Robert White, "Natural Disasters: Acts of God or Results of Human Folly?" in Robert S. White, ed., *Creation in Crisis* (London: SPCK, 2009), pp. 102-21.

[6]J. R. Petit et al., "Climate and Atmospheric History of the Past 420,000 Years from the Vostok Ice Core, Antarctica," *Nature* 399 (1999): 429-36.

[7] Data from National Snow and Ice Data Center, http://nsidc.org.

[8] A useful reference book for those interested in digging deeper into global warming and climate change is John Houghton, *Global Warming—The Complete Briefing*, 4th ed. (Cambridge: Cambridge University Press, 2009).

[9] Mark Aslin and Patrick Austin, "Climate Models at Their Limit?" *Nature* 183 (2012): 183-84.

[10] J. Dukes, "Burning Buried Sunshine: Human Consumption of Ancient Solar Energy," *Climatic Change* 61 (2003): 31-44.

[11] Data from Dr Pieter Tans, NOAA/ESRL (www.esrl.noaa.gov/gmd/ccgg/trends), and Dr Ralph Keeling, Scripps Institution of Oceanography (scrippsco2.ucsd.edu).

[12] United Nations, "International Day for the Preservation of the Ozone Layer: Background," www.un.org/en/events/ozoneday/background.shtml.

[13] Statement from the *Summary for Policymakers of the Synthesis Report of the IPCC Fourth Assessment Report* (2007), p. 1.

[14] Three organizations calculate global average temperature each month. These work independently and use different methods in the way they collect and process data to calculate the global average temperature. Despite this, the results of each are similar. HadCRU curves are from the UK Meteorological Office in collaboration with the Climatic Research Unit at the University of East Anglia, based on around 1.5 million observations each month from land stations, buoys and ships. NASA curves are from the Goddard Institute for Space Studies, part of NASA. NOAA curves are from the National Climatic Data Center, part of the US National Oceanic and Atmospheric Administration.

[15] Redrawn from a record of atmospheric carbon dioxide, methane and temperature extracted from the Vostok Antarctic ice core by Petit et al., "Climate and Atmospheric History of the Past 420,000 Years." The local air temperature is inferred from oxygen isotopes trapped in the ice. Present-day (2012) values of carbon dioxide are from the Mauna Loa observatory, Hawaii, and for methane are the average of mid-latitude northern (Mace Head, Ireland) and southern (Cape Grim, Tasmania) hemisphere readings in 2010.

[16] Redrawn from IPCC, *Climate Change 2007: The Physical Science Basis, Summary for Policymakers* (Cambridge: Cambridge University Press, 2007).

[17] "Public's Priorities for 2010: Economy, Jobs, Terrorism," Pew Research Center, January 25, 2010, www.people-press.org/report/584/policy-priorities-2010.

[18] "Climate scepticism 'on the rise,' BBC poll shows," February 7, 2010, http://news.bbc.co.uk/2/hi/8500443.stm.

[19] Adapted from S. Solomon et al., eds., *Climate Change 2007, The Physical Science Basis, Working Group I Contribution to the Fourth Assessment Report of the IPCC Intergovernmental Panel on Climate Change* (Cambridge: Cambridge University Press, 2007).

[20] Mark Lynas, *Six Degrees: Our Future on a Hotter Planet* (New York: Harper Perennial, 2008).

²¹A review of possible tipping points is provided by Tim Lenton et al., "Tipping Elements in the Earth's Climate System," *Proceedings of the National Academy of Science* 105 (2008): 1786-93.

²²A good book discussing the practicalities of different energy sources for the UK is Donald J. C. Mackay, *Sustainable Energy—Without the Hot Air* (Cambridge: UIT Cambridge Ltd, 2008).

²³See, e.g., *Geoengineering the Environment: Science, Governance and Uncertainty* (London: Royal Society Policy document, October 2009).

²⁴For descriptions of the deliberate efforts to downplay the effects of climate change, see James Hoggan, *Climate Cover-Up: The Crusade to Deny Global Warming* (Vancouver: Greystone Books, 2009); James J. McCarthy, "Climate Science and Its Distortion and Denial by the Misinformation Industry," in Robert S. White, ed., *Creation in Crisis* (London: SPCK, 2009), pp. 34-52; *Smoke, Mirrors and Hot Air: How ExxonMobil Uses Big Tobacco Tactics to Manufacture Uncertainty on Climate Science* (Union of Concerned Scientists, 2007); Naomi Oreskes and Erik Conway, *Merchants of Doubt: How a Handful of Scientists Obscured the Truth on Issues from Tobacco Smoke to Global Warming* (New York: Bloomsbury Publishing, 2011).

²⁵Nicholas Stern, *The Economics of Climate Change: The Stern Review* (Cambridge: Cambridge University Press, 2007). Subsequently in 2008 Lord Stern commented that his report had underestimated the risks of climate change and the damage associated with it, which made the likely costs of inaction even greater than stated in his original report. See David Adam, "I Underestimated the Threat, Says Stern," *The Guardian*, April 18, 2008, www.guardian.co.uk/environment/2008/apr/18/climatechange.carbonemissions.

²⁶Donald A. Hay, "Responding to Climate Change: How Much Should We Discount the Future?" in Robert S. White, ed., *Creation in Crisis* (London: SPCK, 2009), pp. 53-66.

4. WHY HOPE? THE GOSPEL AND THE FUTURE

¹Ian McEwan, "The Day of Judgement," *The Guardian*, May 31, 2008.

²David Orr, "Armageddon Versus Extinction," *Conservation Biology* 19 (2005): 290-92.

³This quote was made famous by Robert F. Kennedy, although it is originally attributed to Harry Lauder and more recently was cited by US President Barack Obama in his 2011 State of the Union Address.

⁴Rowan Williams, "Renewing the Face of the Earth: Human Responsibility and the Environment," Ebor Lecture, York Minster, March 25, 2009.

⁵R. J. Berry, "Sustainability: God's Way or Greenwash?" in R. J. Berry, ed., *When Enough Is Enough: A Christian Framework for Environmental Sustainability* (Nottingham: Apollos, 2007), p. 15.

⁶Kathryn Tanner, "Eschatology Without a Future?" in J. Polkinghorne and M. Welker, eds., *The End of the World and the Ends of God: Science and Theology on Eschatology* (Harrisburg, PA: Trinity Press International, 2000), pp. 222-37.

⁷Stephen Williams, "Thirty Years of Hope: A Generation of Writing on Eschatology," in

K. E. Brower and M. W. Elliott, eds., *"The Reader Must Understand": Eschatology in Bible and Theology* (Nottingham: Apollos, 1997), pp. 243-58.

[8]Karl Barth, *The Epistle to the Romans*, trans. E. C. Hoskyns (Oxford: Oxford University Press, 1933), p. 314. See also Craig C. Hill, *In God's Time: The Bible and the Future* (Grand Rapids: Eerdmans, 2002), pp. 3-11.

[9]Christoph Schwöbel, "The Church as a Cultural Space: Eschatology and Ecclesiology," in Polkinghorne and Welker, *The End of the World*, p. 117. See also Oliver O'Donovan, *Resurrection and the Moral Order: An Outline for Evangelical Ethics* (Grand Rapids: Eerdmans, 1986).

[10]This point has recently been stressed by Michael Bird, *Commentary on Romans* (Grand Rapids: Zondervan, forthcoming).

[11]See the famous Priene Inscription (ninth century B.C.) on the birth of Augustus: "For the world the birthday of the god was the beginning of his good news." The Jewish historian Josephus also notes the proclamations of good news associated with the ascension of Emperor Vespasian (*War*, 4.618; cf. *War*, 4.656).

5. Bringing New Testament Hope Down to Earth

[1]The hymn was written in 1922 by Helen H. Lemmel, with the refrain "Turn your eyes upon Jesus / Look full in His wonderful face / And the things of earth will grow strangely dim / In the light of His glory and grace."

[2]For a compelling and accessible survey of resurrection hope in the Bible and its relevance for Christians today, see N. T. Wright, *Surprised by Hope* (London: SPCK, 2007).

[3]The use of Romans 8 (and Col 1:15-20) as a hermeneutical lens through which to read the rest of the New Testament has also been urged by Cherryl Hunt, "Beyond Anthropocentrism: Towards a Re-reading of Pauline Ethics," *Theology* 112 (2009): 190-98.

[4]In the New Testament see, e.g., Mark 10:6; 13:19; Romans 8:39; Colossians 1:15; Hebrews 9:11; 2 Peter 3:4; Revelation 3:14.

[5]For a detailed argument that Isaiah 24–27 has influenced Paul's thought and language in Romans 8, see Jonathan Moo, "Romans 8.19-22 and Isaiah's Cosmic Covenant," *New Testament Studies* 54 (2008): 74-89.

[6]For a very brief survey of ancient conceptions of how human beings related to their natural environment, see K.-W. Wee, "Environment, Environmental Behavior," in Hubert Cancik and Helmuth Schneider, eds., *Brill's New Pauly: Encyclopedia of the Ancient World*, vol. 4 (Leiden: Brill, 2004), pp. 1002-8.

[7]See, e.g., Mark Lynas, *The God Species: How the Planet Can Survive the Age of Humans* (Washington, DC: National Geographic Society, 2011).

6. Cosmic Catastrophe?

[1]There are also occasions when fascination with the "end times," combined with a misunderstanding of the genre of ancient biblical texts (especially Revelation), leads to an unhappy marriage of both approaches and Christians attempt to discern in current

events biblical portents of impending apocalypse. Such an enterprise mistakes the purpose of a book like Revelation, and we do well to heed Jesus' claim that "about that day or hour no one knows, not even the angels in heaven, nor the Son, but only the Father" (Mk 13:32; Mt 24:36).

[2] Scholars debate whether Peter is actually the author of the letter, but as this question does not affect the authority that 2 Peter has for Christians today, we will not deal with it here. For the sake of convenience, we will refer to the author as "Peter."

[3] See Elizabeth Asmis, "Epicureanism," *Anchor Bible Dictionary*, vol. 2 (New York: Doubleday, 1992), pp. 559-61; Edward Adams, *The Stars Will Fall from Heaven: Cosmic Catastrophe in the New Testament and Its World*, Library of New Testament Studies 347 (London: T & T Clark, 2007), pp. 109-14.

[4] For more details about the text and translation of this verse, see Richard Bauckham, *Jude, 2 Peter*, Word Biblical Commentary 50 (Waco, TX: Word Books, 1983), pp. 303, 316-21; Al Wolters, "Worldview and Textual Criticism in 2 Peter 3:10," *Westminster Theological Journal* 49 (1987): 405-13, esp. pp. 405-8; Peter H. Davids, *The Letters of 2 Peter and Jude*, Pillar New Testament Commentary (Grand Rapids: Eerdmans, 2006), pp. 286-87; and Jonathan Moo, "Continuity, Discontinuity and Hope: The Contribution of New Testament Eschatology to a Distinctively Christian Environmental Ethos," *Tyndale Bulletin* 61 (2010): 31, esp. n. 18. Moo's article also contains material that serves as the basis for this chapter, along with more scholarly discussion and references for some of the arguments that are made in this chapter and at several other points in this book.

[5] This is a point emphasized by Adams, *The Stars Will Fall from Heaven*, p. 226.

[6] See Douglas Moo, "Nature in the New Creation: New Testament Eschatology and the Environment," *Journal of the Evangelical Theological Society* 49 (2006): 449-88, here pp. 466-69.

[7] The earliest certain examples of this usage are from the second century, e.g., Theophilus of Antioch, *Autol.* 1.4; Justin, *2 Apol.* 5.2; *Dial.* 23.2.

[8] This may be how Paul uses the word (Gal 4:3, 9; cf. Col. 2:8, 20).

[9] Recent English-language commentaries that support this interpretation include Bauckham, *Jude, 2 Peter*, pp. 315-16; Steven J. Kraftchick, *Jude / 2 Peter*, Abingdon New Testament Commentary (Nashville: Abingdon, 2002), p. 163; and Davids, *2 Peter and Jude*, pp. 284-87.

[10] Davids, *2 Peter and Jude*, p. 288. Al Wolters makes the point even more strongly: "The text of 2 Pet. 3:10 ... stresses ... the permanence of the created earth, despite the coming judgement" ("Worldview and Textual Criticism," p. 413).

[11] These comments are based on those in Jonathan Moo, "Environmental Unsustainability and a Biblical Vision of the Earth's Future," in Robert S. White, ed., *Creation in Crisis* (SPCK, 2009), pp. 255-70.

7. Jesus, a Thief in the Night and the Kingdom of God

[1] For more on the significance of this title, see the recent collection of essays in Larry W.

Hurtado and Paul L. Owen, eds., *"Who Is This Son of Man?": The Latest Scholarship on a Puzzling Expression of the Historical Jesus* (London: T & T Clark, 2011).

[2]The implied connection between God/Christ and the master in the parable in Luke 12:35-38 is especially evident in the shocking action of the master when he comes and finds his servants ready: rather than the servants waiting on their master, the master waits on them (v. 37)!

[3]We are indebted for this observation and several other insights relating to this parable to the excellent monograph by Matthew S. Rindge, *Jesus' Parable of the Rich Fool: Luke 12:13-34 Among Ancient Conversations on Death and Possessions*, Society of Biblical Literature, Early Christianity and Its Literature 6 (Leiden: Brill, 2012). For readers interested in how Paul takes forward Jesus' teaching and traditional Jewish teaching on the necessity to care for the poor, see Bruce W. Longenecker, *Remember the Poor: Paul, Poverty, and the Greco-Roman World* (Grand Rapids: Eerdmans, 2010), although even this book fails to deal in any detail with some of Paul's most radical claims about the goal of material equality among the people of God (2 Cor 8:13-15).

[4]The most famous version of this critique is the short article by Lynn White Jr., "The Historical Roots of Our Ecologic Crisis," *Science* 155 (1967): 1203-7. For an accessible rejoinder to this sort of critique of Christianity, see Alister McGrath, *The Reenchantment of Nature: The Denial of Religion and the Ecological Crisis* (New York: Doubleday, 2002).

[5]The most useful collection of essays assessing the term and the idea of "stewardship" is R. J. Berry, ed., *Environmental Stewardship: Critical Perspectives—Past and Present* (London: T & T Clark, 2006).

8. Revelation and the Renewal of All Things

[1]For a discussion of the centrality within biblical hope of the resurrection of the body (and its corollary of a material new creation), see N. T. Wright, *Surprised by Hope* (London: SPCK, 2007).

[2]The power of worship to reorient us around our proper center is especially emphasized by Eugene H. Peterson, *Reversed Thunder: The Revelation of John and the Praying Imagination* (New York: HarperOne, 1988). The significance of creation's praise of God in Revelation for our understanding of the natural world and our place within it is developed by Richard Bauckham, *Living with Other Creatures: Green Exegesis and Theology* (Waco, TX: Baylor University Press, 2011), pp. 163-84.

[3]The connection and contrast between Isaiah 6:3 and Revelation 4:8 is developed briefly by Richard Bauckham, *The Theology of the Book of Revelation* (Cambridge: Cambridge University Press, 1993), pp. 46-47.

[4]See Richard Bauckham, *Climax of Prophecy* (London: T & T Clark, 1993), pp. 338-83.

[5]It is sometimes argued that "earth" here refers only to people—e.g., David E. Aune, *Revelation 6–16*, Word Biblical Commentary 52b (Nashville: Thomas Nelson, 1998), p. 645; G. K. Beale, *The Book of Revelation: A Commentary on the Greek Text*, New International Greek Testament Commentary (Grand Rapids: Eerdmans, 1999), pp. 615-16—but John

never uses the word with this limited referent. John's conception is best understood within the context of the prophetic tradition in which human evil is seen as having consequences for all of creation, as in such texts as Isaiah 24–27 and Hosea 4:1-3 (and Rom 8). More immediately, John is alluding to Jeremiah 51:25, where it is Babylon who is destroying "the whole earth."

[6]For a detailed argument that the four living creatures of Revelation are intended as representations of all of animal creation, see Bauckham, *Living with Other Creatures*, pp. 164-80.

[7]Poul F. Guttesen, *Leaning into the Future: The Kingdom of God in the Theology of Jürgen Moltmann and in the Book of Revelation*, Princeton Theological Monograph Series (Eugene, OR: Pickwick, 2009), pp. 128-30.

[8]Ibid., p. 130.

[9]Whatever one may think about the rest of Augustine's classification of the so-called four states of humanity, his description of the redeemed life that applies in the new creation as one where there is not even the possibility of sin and rebellion (*non posse peccare*) comports well with the picture in Revelation 21–22.

[10]For a fuller development of the themes explored in this section, see Jonathan Moo, "The Sea That Is No More: Rev. 21.1 and the Function of Sea Imagery in the Apocalypse of John," *Novum Testamentum* 51 (2009): 148-67.

[11]See Richard Bauckham, "Ecological Hope in Crisis?" *JRI Briefing Paper* 23 (2012), www.jri.org.uk/wp/wp-content/uploads/JRI_23_Hope_Bauckham.pdf.

9. Finding Joy in an Active and Living Hope

[1]In John Stott's final book, *The Radical Disciple: Some Neglected Aspects of Our Calling* (Downers Grove, IL: InterVarsity Press, 2010), he observes that creation care is in fact a necessary (and too often neglected) aspect of what it is to be a radical disciple of Christ.

[2]This point has been made compellingly by Richard Bauckham in a number of articles, chapters and books. See especially Richard Bauckham, *The Bible and Ecology: Rediscovering the Community of Creation* (London: Darton, Longman & Todd, 2010); and Bauckham, *Living with Other Creatures: Green Exegesis and Theology* (Waco, TX: Baylor University Press, 2011).

[3]*The Onion* 46.3 (January 19, 2010), www.theonion.com/articles/how-bad-for-the-environment-can-throwing-away-one,2892.

[4]The NIV supplies the word "all" here, as it is implied in Peter's question and is explicit in the parallel version of this story found in Matthew 19.

[5]We owe this suggestion to Ajith Fernando, from a discussion at Whitworth University, Spokane, Washington, October 3, 2012.

[6]Richard Bauckham discusses the connections between ultimate and proximate hope in "Ecological Hope in Crisis?" *JRI Briefing Paper* 23 (2012), www.jri.org.uk/wp/wp-content/uploads/JRI_23_Hope_Bauckham.pdf.

Scripture Index

Old Testament
Genesis
1, *143, 144*
1–3, *110*
1:2, *160*
1:4, *83*
1:10, *83*
1:12, *83*
1:18, *83*
1:20-27, *159*
1:21, *83*
1:25, *83*
1:26-28, *110, 143*
1:28, *143*
2:7, *159*
2:15, *110, 144*
3, *107, 158*
3:17, *105*
8:11, *121*

Exodus
24:9-10, *98*
32:1-8, *107*

Leviticus
19:18, *142*

Numbers
33:52, *108*

Deuteronomy
6:5, *142*
10:12, *127*
29:17, *108*

1 Kings
22:19, *98*

Job
38:11, *160*
38:26-27, *36*

Psalms
8:1, *22*
8:3-5, *22*
11:4, *98*
36:9, *148*
104:26, *160*
111:10, *139*

Proverbs
1:7, *139*
9:10, *127, 139*
19:23, *127*

Isaiah
6:3, *149*
24, *106, 107*
24–27, *105, 107, 108*
24:4-5, *106*
25:8, *157*
32, *129*
32:15-16, *129*
32:17, *129*
33:9, *52*
34:4, *121*
40:28-31, *162*
61:1-2, *88*
65–66, *129*
66:1, *98*

Jeremiah
12:4, *52*
51:45, *154*

Ezekiel
20:7-8, *108*

Daniel
7, *132, 159*
7:13-14, *132*
7:14, *132*

Hosea
4, *107*
4:1, *107*
4:1-3, *53*
4:2, *107*
4:3, *52, 107*

New Testament
Matthew
5:12, *99*
5:34, *98*
6:19-20, *99*
6:25-34, *139*
10:39, *168*
11:28-30, *168*
16:25, *168*
19:28, *168*
22:37-39, *142*
24:36, *133*
24:42-51, *133*
24:43, *119*
25:40, *54*
28:20, *144, 163*

Mark
12:30-31, *142*
13:32, *133*
13:33-37, *133*

Luke
4:18-21, *88*
5:24, *132*
6:10, *88*
7:16, *88*
7:19, *88*
7:22-23, *88*
7:34, *132*
9:22, *132*
9:26, *132*
9:44, *132*
10:27, *142*
12, *119, 132, 133, 134, 136, 139, 142, 145*
12:4, *139*
12:5, *139*
12:6, *139*
12:7, *139*
12:11-12, *144*
12:13, *137*
12:13-21, *140*
12:15, *137*
12:16, *136*
12:17-19, *136*
12:21, *136*
12:22-34, *139, 140*
12:24, *140*
12:31, *140*
12:32, *80, 145*
12:33, *140, 141*
12:35-48, *139*
12:37, *133*
12:38, *134*
12:39, *119, 132*
12:39-40, *132*
12:40, *134*
12:41, *135*
12:42, *135, 142*
12:42-46, *132*
12:42-48, *143*
12:44, *135*
12:48, *135*
13:12-16, *88*
17:20-21, *134*
17:24, *134*
18:1, *164*
18:7, *163*
18:8, *164*
18:22-23, *167*
18:25, *167*
18:28, *167*
18:29-30, *168*
18:31, *132*
21:27, *132*
22:22, *132*
22:48, *132*
24:7, *132*
24:46-47, *89*

John
5:17, *144*
11:35, *169*
14:1-4, *99*
18:36, *163*

Acts
1:11, *133*
3:18-21, *89*
3:19-21, *168*
4:10-12, *89*
9:1-19, *102*
9:31, *127*
17, *91*
17:24-26, *91*
17:26-28, *91*
17:30, *91*
17:31, *91*
17:32-34, *91*
24:25, *126*

Romans
1:2, *90*
1:2-6, *90*
1:21-23, *107*
4:17, *83*

8, *98, 100, 101, 102, 105,*
106, 107, 109, 110,
111, 112, 113, 114,
116, 117, 118, 144
8:1, *102*
8:3, *102*
8:12-17, *112*
8:14-17, *102*
8:18, *101*
8:18-24, *101*
8:18-25, *103*
8:19-22, *102*
8:19-23, *102*
8:20, *110*
8:20-21, *105*
8:21, *105*
8:21-22, *52*
8:23, *101, 102*
8:28, *101, 102*
8:29, *102*
12:15, *169*

1 Corinthians
3:10-14, *123*
3:13, *123*
4:16, *111*
6:20, *112*
8:4, *108*
12:2, *108*
13, *96*
13:12, *96*
13:13, *19, 82*
15, *105, 112*
15:3, *90*
15:3-5, *90*
15:4, *90*
15:20-28, *92*
15:22, *102*
15:58, *125*

2 Corinthians
4:4-5, *92*
4:18, *99*
5:1-4, *99*
5:4, *105*
5:17, *112*
9:13, *93*
11:24-28, *101*
12:2-4, *99*

Galatians
1:11-24, *102*

Ephesians
2:6, *99*
5:5, *108*

Philippians
1:23, *100*
1:27, *86*
2:1-18, *111*
2:12-13, *127*
3:20, *99, 100*

Colossians
1:15, *144*
1:16, *92*
1:19-20, *92*
1:20, *112*
3:5, *108*
3:10, *144*

1 Thessalonians
1:9, *108*
4:13, *169*
5:2, *119*
5:4, *119*

2 Timothy
1:10, *92*

Hebrews
2, *92*
2:14-15, *92*
8:1, *98*
11:3, *83*
11:13-16, *99*

James
2:8, *142*
3:1, *135*
4:13, *137*
4:14, *137*

1 Peter
1:4, *99*
1:7, *123*
3:14, *127*
3:22, *120*

2 Peter
1, *127*
1:1, *124*
1:3, *126*
1:5-8, *126*

1:8, *128*
1:8-11, *124*
1:18, *120*
2:14, *117*
2:19, *124*
2:19-20, *117*
2:21, *124*
3, *115, 116, 117, 118,*
119, 120, 121, 123
3:3-4, *117*
3:5-7, *118*
3:6, *121*
3:7, *122*
3:8-9, *118*
3:9, *124*
3:10, *118, 119, 121,*
131
3:10-13, *116, 117, 118*
3:11, *124, 125, 126*
3:11-12, *125*
3:12, *120*
3:13, *123, 126*
3:13-14, *129*
3:14, *129*
3:15-16, *116*
3:17, *124*

1 John
4:18, *127, 165*

Revelation
1:5, *148*
3:3, *119*
4–5, *156*
4:1, *99*
4:8, *149*
5:4, *161*
5:6-10, *149*
5:9-10, *147*
5:10, *143, 157, 161*
5:13, *147, 159*
6, *156*
6–16, *149*
8:9, *159*
10:6, *159*
11, *157*
11:7, *159*
11:15, *147, 156*
11:15-18, *155*
11:17, *157*
11:18, *155, 156*
12, *156*

12:5, *156*
12:10, *149, 156*
12:10-11, *149*
12:12-17, *149*
13, *149*
13:1, *159*
13:3, *153*
13:4, *153*
14:7, *159*
16:10, *159*
16:15, *119, 123*
17, *150*
17–18, *151*
17–19, *149*
17:3-6, *153*
17:6, *153*
17:8, *159*
17:9, *151*
18, *150, 152*
18:3, *152*
18:4, *153*
18:7, *152*
18:11-13, *152*
18:13, *152*
18:19, *152*
18:20, *154*
18:21, *159*
18:23, *152, 153*
18:24, *152*
19, *150, 154*
19:1, *155*
19:1-2, *155*
19:1-10, *155*
19:2, *153*
19:3, *155*
19:4, *155*
19:6, *155, 156*
19:8, *123, 124*
20:10, *149*
21–22, *111, 129, 157*
21:1, *149, 157, 158*
21:2, *97, 100, 157*
21:5, *113, 157, 158,*
168
21:22, *157*
21:24, *157*
22, *158*
22:2, *158*
22:3, *157*
22:5, *143*